T0320947

Transportation Systems Security

Transportation Systems Security

Allan McDougall, BA, BMASc, PCIP, CMAS
Robert Radvanovsky, CIFI, CISM, REM, CIPS

Dr. Carl Clavadetscher
Technical Editor

CRC Press
Taylor & Francis Group
Boca Raton   London   New York

CRC Press is an imprint of the
Taylor & Francis Group, an **informa** business

CRC Press
Taylor & Francis Group
6000 Broken Sound Parkway NW, Suite 300
Boca Raton, FL 33487-2742

© 2008 by Taylor & Francis Group, LLC
CRC Press is an imprint of Taylor & Francis Group, an Informa business

No claim to original U.S. Government works
Printed in the United States of America on acid-free paper
10 9 8 7 6 5 4 3 2 1

International Standard Book Number-13: 978-1-4200-6378-3 (Hardcover)

Library of Congress Cataloging-in-Publication Data

McDougall, Allan.
    Transportation systems security / authors, Allan McDougall, Robert
Radvanovsky.
        p. cm.
    Includes bibliographical references and index.
    ISBN 978-1-4200-6378-3 (hardback : alk. paper)
    1. Transportation--Security measures--United States. 2.
Transportation--United States--Safety measures. 3. National security--United
States. I. Radvanovsky, Robert. II. Title.

    HE194.5.U6M43 2008
    363.12'060973--dc22   2008010872

Visit the Taylor & Francis Web site at
http://www.taylorandfrancis.com

and the CRC Press Web site at
http://www.crcpress.com

Dedication

First, and foremost, my thanks to Angela, who has helped me keep things in balance throughout this effort. And thanks to all those who still stand on the walls and watch our world (physically and logically) so that we have this opportunity to write and read at our leisure.

A. M.

To my wife, Tammy, whom I love dearly, for being my sounding board and biggest fan; and to Dr. Carl Clavadetscher, who was inspirational to me early in life and has recently resumed his role as my friend and mentor—thank you.

R. R.

Everything should be made as simple as possible, but not one bit simpler.

Albert Einstein

Contents

3   Business Goals and Mission Analysis 33

4   General Definitions and Approaches 53

11   Establishing and Monitoring Learning Systems..... 183

12   Fragility and Fragility Analysis Management 197

Preface

This book is broken into 12 categories representing a gradual build of knowledge and ideas exchange. These concepts represent months, if not years, of research, as well as strategic and tactical development, which emphasize those concepts comprising functions within and throughout the transportation sector.

These concepts are categorically subdivided into unique and prioritized levels, beginning with Chapter 1, as the broadest chapter of the book, Chapter 2 slightly less broadened than Chapter 1, and so forth. Each subsequent chapter builds from the previous one, emphasizing a different meaning conveyed so it can be structured and remembered in an easy, cognitive fashion.

Chapter 1, "Introduction to Transportation Systems," provides the basis for the entire book: what is relevant about transportation systems and why they are important to countries throughout the world. The chapter contains some terms and definitions covering a brief synopsis of the intent of the book and what is to be expected from management and critical infrastructure protection (CIP) security professionals to work and operate within the transportation sector.

Chapter 2, "Transportation System Topology," describes both physical and information systems, their layout, and what is being considered within the transportation system. Chapter 3, "Business Goals and Mission Analysis," gives the core considerations when looking at performance considerations within the transportation system. The mission analysis is outlined, along with who controls what parts of the process, how far control is extended, and who oversees controls within the system, while introducing the concepts of capacity and system performance within the transportation system.

Chapter 4, "General Definitions and Approaches," provides definitions used throughout the book and also introduces some newer, or more refined, terms and their concepts to these definitions. Chapter 5, "Local versus System Approaches," discusses the concept of "Follow the Pipe"—that threats, assets, vulnerabilities, safeguards, and so forth may not be under the control of the local system—thus, the statement that a more defined "domain-based" approach may be needed or considered.

Chapter 6, "Criticality, Impact, Consequence, and Internal/External Distributed Risk," introduces the concept of value for personnel, assets, facilities, information (and their activities) and so forth in terms of how they might support performance throughout the transportation system. This chapter primarily explains the terms *criticality, impact,* and *consequence.* Chapter 7, "Mitigation and Cost Benefit," establishes the idea of linking security measures with the performance of the entity in terms of critical path activities so that security may be seen as a value-added benefit and not so much as a brake upon performance.

Chapter 8, "Certification Accreditation, Registration, and Licensing," puts forward the concept of certification and accreditation as litmus for whether or not an entity can perform within the system. A regulatory body incorporates oversight of the certification and accreditation process. Chapter 9, "Continuity of Operations," details the concept of continuity of operations planning (COOP), compares it to business continuity planning (BCP) concepts, and shows how critically important and vital these concepts are to contingency planning at all levels within the system and to any circumstance, disaster, and so forth. The chapter contrasts and provides a comparative analysis between COOP and BCP.

The idea of Chapter 10, "Networks and Communities of Trust," is to provide for safeguards to layer outward from the entity and cross the system. These networks would include abilities to prevent, detect, notify, and respond to and recover from situations, events, and disasters. Chapter 11, "Establishing and Monitoring Learning Systems," discusses core characteristics associated with sharing of information and methods within trusted communities inside the transportation system such that the system may take advantage both from lessons learned as well as best business practices. The focus is communities of trust and their practices, which includes identification and normalization of security practices within an organization such that security becomes an integrated component within the system, not just imposed on top of it.

Probably the most important chapter of the entire book is Chapter 12, "Fragility and Fragility Analysis Management," focusing on the latest concept of capacity assurance and takes a holistic approach to the transportation infrastructure rather than focusing on its individual components. This chapter introduces a new term called *fragility* and the measurable factors involved in determining the fragileness of an organization.

This publication offers an aid in maintaining professional competence, with the understanding that the authors, editor, and publisher are not rendering any legal, financial, or any other professional advice.

Due to the rapidly changing nature of the homeland security industry, the information contained within this publication may become outdated; therefore, the reader using this book should consider researching for alternative or other professional or more current sources of authoritative information. The significant portion of our text was based on research conducted over the past year (2007) from several government resources, publications, and Internet-accessible Web sites, some of which may no longer be publicly available or may have been restricted due to laws enacted by a particular country's federal or national government.

The views and positions taken in this book represent the considered judgment of the authors. They acknowledge, with gratitude, any inputs provided and resources offered that contributed to this book. Moreover, for those who have contributed to the book's strengths and its characteristics, both of us would like to say thank you for your contributions and efforts. For any inconsistencies (which should be very few) that have been found, we alone share and accept responsibility for them and will gladly make corrections as needed.

Acknowledgments

This work can really be described as a combination of three factors in what can only be described as the "perfect storm." At the beginning of my career, two people—Claude Pilon and Ron Newman—set the stage for what can now be considered more of a calling than a career. Claude, in his odd balance of manager and athlete, rekindled that odd desire to achieve "one more inch" as he guided me through a broad range of security expertise. Ron, on the other hand, managed to show the sheer joy that the security field could bring because of hard work and the knowledge that work done has contributed to the success of an organization.

The second part of this storm is the maelstrom of people, both inside and outside government, who have helped me form ideas as I work through many of the challenges in this field. These people brought ideas from Public Safety, Communications Security Establishment, the Royal Canadian Mounted Police, Canadian Security Intelligence Service, Transport Canada, Customs, Immigration, the Department of Fisheries and Oceans, the Canadian Coast Guard, the U.S. Coast Guard, the Canadian Forces, and a host of private-sector firms and independent consultants that each had a part to play in how to approach this challenge. Whether these people were working at the national headquarters of their respective organizations or in the industry, I have to say that the challenges we faced together have helped mold the "how" I use to approach this challenge.

Finally, there are those that helped me work toward actually communicating these ideas. The first of these is my coauthor, Bob, who helped me try to pull the statue out of the stone. I can honestly say that working on this with him has been one of the more challenging and rewarding experiences I have had to date. The second is a former colleague from the Canadian Forces, Richard, who helped with making sure that I kept things practical and did not digress into the wildly theoretical. Finally, at the end of the day, there is my wife Angela, who had to look at much of this and tell me, without bruising a fragile ego, whether things actually made sense at the end of the work.

At the end of this, I can only present this work as a humble offering with the hope that it will help others challenge their own thoughts and assumptions with the same success that I have had.

A. M.

Aside from my wife, Tammy, my former college professor, mentor, and friend Dr. Carl Clavadetscher has been a good comrade by agreeing to provide his expertise as technical editor of this book. His insightfulness on present and current homeland security-related topics has been invaluable. Both Allan and I believe that this can provide a fundamental understanding as to how the transportation system is perceived holistically—in its entirety—thus ensuring that goods, services, and people are safely and accurately delivered to their destinations in a timely manner. We believe that our modified perception of how the transportation system paradigm operates will enhance and enable CIP security professionals worldwide, not only by improving their business operations but also by altering their reality to an even larger arena containing all aspects of all modes of transportation. Ironically, many of the concepts within this book can be "cookie-cut" into other venues of CIP, to which Dr. Clavadetscher has provided substantial guidance. We believe that this is only the beginning of future publications on this exciting and (understandably) complex topic, as it will only gain more attention as time progresses.

Last, but not least, a special thanks to my coauthor, Allan, who provided me with this wonderful opportunity to continue publishing my ideas and concepts pertaining to this exciting topic. It is a rather weird yet satisfying feeling to find someone who has a similar—perhaps even as warped—a mind as my own. Thank you, Allan, for encouraging me not to be confined by the box.

R. R.

Authors

Allan McDougall is co-director of Evolutionary Security Management, an independent research and consulting firm focusing on critical infrastructure assurance, critical infrastructure protection, and security management. His experience includes service within Canada's Armed Forces as a combat engineer before entering public service. His experience within public service included a range of security-related positions that spanned policy, threat analysis and risk assessment, physical security, information security and personnel security screening related to the transportation sector, often within lead agencies.

Allan is a graduate of both the University of Western Ontario (London, Ontario) and the Royal Military College of Canada (Kingston, Ontario). Additionally, he holds a number of certifications from professional organizations associated with emergency preparedness, emergency response and critical infrastructure protection in Canada and the United States. His dedication to continuous learning has branched into providing training associated with transportation security (including recognized training in the marine security domain) and his holding the chair of the Anti-Terrorism Accreditation Board's Transportation Security Committee.

Bob Radvanovsky is an active professional in the United States with extensive knowledge in security and risk management, as well as business continuity and disaster recovery planning and remediation. He has a Master's of Science degree in computer science from DePaul University in Chicago. In addition, he has been awarded several professional certifications in the fields of information technology and security, including that of Certified Information Forensics Investigator for specialization in criminal IT forensics management.

Bob has special interests and knowledge in matters of critical infrastructures, and has published a number of articles and whitepapers on the topic. He has been significantly involved in establishing security training and awareness programs through his company, Infracritical. As part of this effort, he is involved with the Department of Justice INFRAGARD program, as well as several professional accreditation and education institutions, specifically on the topic of homeland security and critical infrastructure protection. Bob's first book, entitled *Critical Infrastructure: Homeland Security and Emergency*

Preparedness was a reference work dealing with emergency management and preparedness, defining what is "critical infrastructure protection."

Both authors have established a private research enterprise investigating concepts surrounding critical infrastructures, how they are impacted, and how they may be preserved.

Introduction to Transportation Systems

1

Objective: At the end of this chapter, the reader will:

- Understand that the Transportation System operates beyond traditionally accepted definitions in terms of nodes, boundaries; and,
- Understand that securing the system means that any capacity associated with the system should be discussed.

1.1 Introduction

The nation's critical infrastructures and key resources—including those cyber and physical assets essential to national security, national economic security, and national public health and safety—have been and continue to be vulnerable to a wide variety of threats. In 2005, Hurricane Katrina devastated United States' Gulf Coast,[1] damaging critical infrastructure such as oil platforms, pipelines, and refineries, water mains, electric power lines, and cellular phone towers. The chaos resulting from this infrastructure damage disrupted the functioning of government and business alike and produced cascading effects far beyond the physical location of the storm and even across North America. A series of coordinated bombings in 2005 that struck London's public transportation system demonstrated how an attack on this system could disrupt a city's transportation and mobile telecommunications infrastructure. These concerns outline a number of documents within the United States and Canada, including, but not limited to, the Presidential Commission on Critical Infrastructure Protection[2] and the National Critical Infrastructure Action Plan[3] activities within Canada.

Historically, the transportation sector, as one of the more visible sectors, is a frequent target for terrorist attacks. Recent incidents over the past several years have reinforced this observation. Within the transportation sector, two main activities are undertaken: (1) the movement of people; and (2) the movement of goods (i.e., cargo or freight). In both cases, the mission of the transportation sector is to ensure that persons or goods are moved to the right location within the right timeframe and arrive in an acceptable condition. Where the focus is on moving people, such as with the airline and ferry industries, the mission is to ensure that passengers arrive on time and in good health (or at least the health they started the voyage on) at their intended destination without being subjected to undue risks.[4] The focus is about moving goods, ensuring that goods arrive in acceptable condition or with only acceptable levels of loss at their intended destination. This supports such systems as the "just-in-time" supply system and other business activities.[5] These primary interests must be balanced with the matter of those working within the system since the major concern about transportation security often is cargo security, which can pose a serious threat to many cargo handlers, facilities, and the ports (or terminals) where cargo is exchanged.[6]

1.2 Requirements for Securing the Sector

Participation and cooperation of both public and private organizations is imperative and must be obtained to successfully apply safeguards against threats to transportation facilities. This is a key first step to securing the transportation sector as these safeguards provide services that enable the sector its capacity. These players do not necessarily fall under the same administrative or legal constraints. Consider, for example, the civil aviation and marine industries that would have significant interaction with the following:

- International organizations, such as the United Nations (UN) International Maritime Organization (IMO)[7] and International Civil Aviation Organization (ICAO)[8]
- Federal agencies responsible for transportation safety and security (e.g., U.S. Coast Guard, Transport Canada, Transportation Security Administration)[9]
- Law enforcement (federal, state/provincial/territorial, and local-level responsibilities)
- Emergency management agencies and first-responder organizations
- Transit and port authorities

- Private transportation provider organizations
- Providers of the necessary goods and services within the port community, including contractors and labor organizations

Among other things, the cooperation among all of these groups attempts to address the following elements:

- Physical security of terminals (e.g., airports, railheads, ports)
- Physical security of routes between those terminals (e.g., rail lines, straits, commonly used flight paths)
- Operations security in terms of the trust in processes and in the effort to avoid missed steps (e.g., misplaced shipments) and to ensure predictable and acceptable levels of performance
- Public safety (in terms of the reduction of impacts and consequences to communities that have, throughout history, clustered around this infrastructure)
- Security of communications and information systems facilities (e.g., systems associated with the ability to locate and coordinate the movement of assets and personnel at all levels of the process)
- Development and dissemination of information regarding incidents (particularly the ability of management systems at all levels to appropriately communicate lessons learned and best business practices in a competitive environment)
- Conduct assessments against potential threats involving transportation facilities or its operators (e.g., threats and kinds of threats that are related but that combine to generate new avenues of risk, such as narcotic-sponsored terrorism)

When looking at the transportation sector, one also has to understand the fine balance required, particularly at the management level, to allow potentially conflicting priorities. The first of these priorities deals with impacts primarily associated with loss of infrastructure—caused by all kinds of disasters. Balanced against this, however, are the impacts associated with the loss of performance in an increasingly competitive world where the ability to compete largely determines the position of trading partners and nations on the world stage—represented by efforts such as the traditional business continuity planning and more recent commerce resumption planning. When looking at the transportation sector, the focus shifts from protecting infrastructure for its own sake to protecting that infrastructure based on the capacity it brings into the overall system. This protection of the capacity still involves dealing with many of the traditional threats and hazards (e.g., terrorism, organized crime) but improves the overall system so that the channels through which movement occurs are enhanced in such a way as to be able to adapt—rapidly—to any disruptions, accidental or otherwise, manmade or natural.

1.3 The Transportation Sector as Linked Systems

The transportation sector is one of the larger and more unusual sectors in that it consists of a complex set of systems of systems. Confusing as it may sound, each transportation system is dependent, either partially—or completely—on another transportation system. What makes the transportation sector so difficult to secure is not so much tackling methods of securing one system but all of them at the same time, due to their interconnectivity with one another. In this aspect, one might argue that the transportation system, instead of being viewed as an individual collection of personnel, assets, facilities, information, and activities operating in isolation and assessed in terms of individuals values, must be looked at in the context of a living organism subject to the same Darwinian laws applicable to any organism.

Unlike the organism, however, the management of this system is not centrally controlled but is accomplished through a distributed system. The management of this system at tactical (facility), operational (regions), modal (maritime, rail, aviation, surface), and strategic (system, international) levels poses its own challenges. Traditional management structures work within most organizational silos or stovepipes (as defined by their administrative responsibilities), taking into account liabilities resulting from impacts/consequences outside of the same. Management of the transportation system requires a carefully coordinated and collaborative approach that focuses on the ability of clear communication of key information across normally competitive lines for the benefit of the overall system.

To list all of these systems would be a large task of itself; however, Table 1.1 gives a summary of the various transportation systems that make up the transportation sector. Cargo may be differentiated by several subelements, which includes hazardous (chemical, biological, radiological, nuclear, environmental) versus nonhazardous (unprocessed goods) materials transport and modal (containerized freight) versus nonmodal (raw goods, such as aggregate rock, or finished manufactured products). One of the major challenges, however, in the transportation sector is being able to identify when categorization is necessary and when it is less than relevant to the issues at hand.

1.4 Impact Resulting from System Failure or Interruption

The destruction of a major roadway, highway, rail line, port, or airport could cease or severely limit the flow of goods and services in and out of the region.

Table 1.1   Transportation Sector Systems

Cargo Moved	System	Transport Type	Examples
Persons	Mass transit	Bus	Municipal bus/transit
		Light rail	Light rail
		Commuter rail	Greater Toronto Transit Authority (GO Transit), Chicago's Metra, Washington Metropolitan Transit Authority (WMTA), and Maryland Rail Commuter (MARC; part of the Maryland Transit Authority)
		Ferries	The CAT, British Columbia ferries
		Taxi/limo	Various local
	Passenger	Bus	Greyhound
		Rail	Amtrak, VIA Rail Canada
		Air	American Airlines, Air Canada
		Marine	Cruise and passenger
Objects	Cargo/freight/bulk	Surface	United and other trucking
		Rail	Canadian Pacific, Union Pacific
		Maritime	Hapag Lloyd, COSCO Group
	Package/cargo/freight	Air	FedEx, United Parcel Service (UPS)
		Package	U.S. Postal Service (USPS), Canada Post

This could result in potentially catastrophic losses to the economy, not to mention health and welfare, depending on the type of services inhibited.[10] Failures from other sectors can have potentially significant impacts on the transportation sector; for example, an area-wide loss of power to traffic control systems may lead to increased number of occurrences of accidents involving both vehicles and pedestrians as well as increased traffic congestion that could interfere with emergency response and recovery efforts. More obviously, a disruption in the production of fuel could also lead to a situation where the components within the transportation sector could cease to function.[11] Local and regional highway and road systems would also become chaotic if the county and its municipalities lose the ability to monitor traffic flow, equipment status, and closed-circuit cameras that regulate the steady flow of goods and people. Loss of public transportation services (e.g., subways, commuter rail, buses and taxis) would affect the ability of hundreds of thousands—or depending on the level of the impact (local versus regional versus national) perhaps even millions—of users of the system to get to work, shop, school, or medical appointments. Disruption of a rail service would cause significant capacity problems within the transportation system, possibly interrupting the supply of vital resources necessary to the health and safety of its citizens. Significant disruptions of functions at ports of entry (airports and maritime ports) might quickly cut the area off from supplies, food, people, and commerce. These interruptions could have a debilitating effect on the economic health of a given region as well as greatly impact human health and safety in the event of an emergency.[12] If one is to consider the image of the "transportation organism," many might relate this in terms of unacceptable levels of impact or consequence being able to spread through the system like an infection. One infected, or exploited, point in the system then infects other parts of the system, and this infection could be carried throughout the system until the system's defenses are overwhelmed by it. For example, when one looks at containers, the fact that a container could move by truck (factory to rail), rail (railhead to port), and then ship (port to port) means that vulnerabilities within one sector can be used to create an impact in another sector. Efforts in supply chain security and supply chain management are currently grappling with this challenge.

1.5 Trends within the Transportation Sector

The number of attacks against mass-transit systems appears to be increasing worldwide, partially because (again) transportation systems are one of the more visible targets for individuals or organizations of nefarious intent. Attacks on transportation systems, especially mass-transit and commuter systems, will have the greatest psychological impact in terms of loss of human life; however, maritime ports of entry have the potential to have an even larger of impact—if affected. On one hand, urban centers and their reliance on mass transit to relieve congestion on the city's highways will continue to drive the potential impact upward in terms of both the number of potential people affected directly (i.e., killed or injured) and the perceived inescapability of fear by having to continue to use a system that has been attacked (i.e., public confidence). On the other hand, one must take note of the increased volume of global goods and services exchanged between countries via shipping. Consider, for example, the Port of Vancouver, which in 2001 handled 1.1476 million containers, whereas in 2005—only four years later—it handled 1.767 million containers.[13] Consider also the fact that with the potential increase of trade with the Far East, particularly with agreements signed by China and Pakistan in terms of economic cooperation, as this number is anticipated to increase significantly over the next several years. Consequently, a disruption within the transportation sector may impact for other sectors or its own if one only looks within the sector itself.

Like the human body, however, things must be taken in balance. The transportation sector is the bloodstream carrying oxygen and nutrients throughout the body. It can either facilitate an attack against another sector (by carrying the hostile or threatening agent to its target) or can actually cause impacts into other parts of the system by failing to meet its mission to ensure the supply of critical nutrients and oxygen (comparable to, e.g., strokes). The various interdependencies of sectors on each other can play crucial roles when one looks at the basic goal of health or, in business terms, the financial health and prosperity of an organization. The magnitude of consequences of each event may be rising (more casualties, more destruction, greater or more significant impacts on more than one sector), partially due to our growing dependence on other sectors, or the volume of the flow of goods and services that are increasing within the sector. Either way, the risk of a catastrophic event is increasing.

1.6 Fragility and Reliability

The challenge for those within the transportation sector is to avoid such events by planning their activities through reliable routes. This poses a significant challenge for the industry. First, any points that demonstrate a propensity to failure (fragility) must be identified, and their location must be communicated back into the industry. These points of failure represent operational risk to those working within the industry.

The method of identifying and communicating these possible points of failure must be consistent, or at least comparable, across the system if the system-level approach is to be taken. Failing to recognize this can lead to errors in translation at any step of the process. These errors in translation can then fail to convey the appropriate message to those who need to receive it, leaving the system blindly at risk.

In looking for this common approach, the question is whether or not fragility represents a new science, an evolution of an existing science, or a qualitative construct that will eventually become a science. One science, albeit based on probability studies, is reliability engineering, which describes the probability "that a system will perform its intended function over a period of time within stated constraints."[14] For individual facilities and infrastructure entities, this provides one avenue or approach to the concept of fragility. Linking this evolving science and its relatively established doctrine provides one additional option.

The importance of this operates on a personal level for each of us. Referring back to the example using public transit, we each want to have an understanding of whether or not the bus or train on which we are embarking will get us to our destination safely and on time. The challenge is that the bus or train does not operate within an isolated environment, and we may even require transfers to other buses or trains to reach our destination. The consideration applied to each individual piece of infrastructure does not reflect the whole picture.

The transportation sector is made up of a network of these individual, but interconnected (and often interdependent), pieces of infrastructure. The impacts of one piece of this type of infrastructure tend to influence the operations of other pieces of infrastructure. The goal for planners is to assure that capacity offered through this collective can actually meet its demand (or even its planned demands).

The next layer in fragility defines how the network reacts to the loss of these individual components. In some cases, the infrastructure may not have a significant influence on the overall performance of the regional network. But what happens if that infrastructure provides a key service or delivers the majority of a certain kind of service within the region? The impact associated with that kind of event makes it all the more disruptive as these impacts move throughout the network.

In addition to its relationship to reliability (at the local level), fragility relates at the network level to the propensity of that system to fragment and eventually dissolve. A number of factors will play a role in the nature, rate, and extent of this fragmentation.

1.7 Understanding Transportation System Security

Transportation system security may be described in terms of two efforts. The first effort involves the ability to protect the infrastructure itself. This infrastructure provides services to the overall system that generates its capacity. Thus, a level of priority has to be given to certain kinds of infrastructure. This includes infrastructure that provides the highest capacity (or even an overwhelming capacity) or a unique but required input into the overall system.

Transportation system security resembles a number of efforts that can be found within the force protection, corporate security, business continuity planning, or continuity of operations planning realms or domains. This effort is also the focus of many regulatory efforts that hone in on the facility or entity demanding a level of compliance. Each of these is involved, in some sense, with putting in place necessary safeguards to ensure that the infrastructure can perform as intended.

This security role only addresses the transportation system security from the local (tactical) level. Again, the mission of the transportation system is to be able to move persons and goods between points—nothing more. While a point-to-point system would be relatively simple to administer, it is also relatively rare. What the transportation system embodies is a collection of these point-to-point systems that must be able to interact in a coordinated way if they are to accomplish their goals.

In this context, transportation system security delves into something that might be described in terms of protecting the capacity of the overall system. Today, intelligent traffic systems and similar activities communicate disruptions within the system, allowing for planners to route around those disruptions. This is done to try to limit the impact's ability to spread throughout the system, much in the same way that the body would encapsulate a foreign growth. While today's goal is progressing, the ultimate goal is to be able to reliably predict fragility within the system and therefore to foresee the problems before they reach catastrophic proportions.

Notes

1. Hurricane Katrina, which struck the Gulf Coast on Aug. 29, 2005, was the most destructive—and costly—natural disaster in U.S. history. Though the unprecedented magnitude of the storm slowed the initial response, the year following was highlighted by an equally unprecedented recovery effort. In the immediate aftermath of Katrina, nearly 275,000 Gulf Coast residents were forced into congregate (group) shelters. At its peak in October 2005, some 85,000 families were being provided transitional housing with hotel or motel rooms in more than 40 states. FEMA, "Hurricane Katrina, One-Year Later," http://www.fema.gov/news/newsrelease.fema?id=29108.
2. The President's Commission on Critical Infrastructure Protection (PCCIP) identified information sharing as "the most immediate need" for the development of enhanced infrastructure assurance. The PCCIP called for "the creation of a trusted environment that would allow the government and the private sector to share sensitive information openly and voluntarily" and recognized that "success [would] depend on the ability to protect as well as disseminate needed information." Its objective was to facilitate the flow of information three ways: among government entities, among private-sector entities, and between government and the private sector. To accomplish this, the commission recommended creation of various information-sharing structures within government and the private sector and also identified a number of potential legal "impediments" to information sharing. "NISSC Panel Proposal," http://www.csrc.nist.gov/nissc/1999/proceeding/papers/o06.pdf.
3. Public Safety Canada, "Creation of the NCIAP," http://www.publicsafety.gc.ca/prg/em/nciap/creation-en.asp; Public Safety Canada, "National Critical Infrastructure Assurance Program: Assessment of Canada's National Critical Infrastructure Sectors," http://ww3.ps-sp.gc.ca/critical/nciap/nci_sector1_e.asp.
4. This was illustrated following the attacks of 9/11, which, while significant, took the airlines months to recover from. Following those attacks, governments were also required to take significant steps with respect to the security of the airline industry, something that is remarked on in reports from the Government

Accountability Office in the United States and in various Senate reports, including the recent Senate Committees on National Security and Defense (SCONSAD) released in March 2007.

5. This concern is reflected in reports produced by such organizations as the Andersen Economic Group, which published its report on the 2002 West Coast Port shutdown at a cost of approximately $1.67 billion for a period of 12 days. The report is available through the Andersen Economic Group's website. Recently, this was also illustrated in the General Motors Strike (September 2006). The strike in the United States quickly stopped the flow of components into Canada, which led to the closure of two plants in Ontario because those critical components were not available. This lack of flexibility in the supply chain raises a significant question as to how to balance the need for some level of flexibility with cost efficiency.

6. U.S. Department of Transportation, Research and Innovative Technology Administration, "National Transportation Technology Plan: 8: Transportation Infrastructure Assurance," http://www.volpe.dot.gov/infosrc/strtplns/nstc/nttplan/chap8.html.

7. http://www.imo.org.

8. http://www.icao.int.

9. Within the maritime industry, for example, the UN houses the International Maritime Organization, which was in the news in 2004 as being the entity responsible for the International Ship and Port Facility Security (ISPS) Code. This code is not actually a law but is rather a series of amendments to the Safety of Life at Sea (SOLAS) Convention. Nations, to remain part of a more trusted network, have to agree to abide by these conventions but are then required to enshrine them into their own legislative frameworks, such as the Canada's Marine Transportation Security Act and the U.S. law with the same name. Under the convention, however, nations are required to identify a designated authority that serves as the point of contact and a lead authority within that government that will meet the obligations defined within the convention.

10. Port strikes on the west coast of Canada and the United States have illustrated this principle with grain shipments being delayed and eventually halted.

11. In February 2007, the fire at the Nanticoke Imperial Oil Refinery led to a reduction in the available fuel supply in Ontario. This disruption, exacerbated by a disruption in rail service, resulted in the movement of goods by truck being slowed because of long lines at fuel pumps.

12. "Washington State Homeland Security, Region 6, Critical Infrastructure Protection Plan, Appendix 5: Transportation Sector," September 2005, http://www.metrokc.gov/prepare/docs/Region6CIP/Region%206%20CIP%20Basic%20Plan%20-%20September%202005.pdf.

13. Ibid.

14. Port Vancouver, "2005 Statistical Summary," December 9, 2006, http://www portvancouver.com/statistics/2005_statistical.html.

15. Drawn from "Reliability Engineering," http://en.wikipedia.org/wiki/Single_point_of_failure.

Transportation System Topography

2

Objective: At the end of this chapter, the reader will:

- How the transportation system is laid out
- That the system relies on supporting modal networks that function as systems
- How demand affects the transportation system

2.1 Introduction

Most people view the transportation system as moving between two terminals using a single mode. When an individual wants to commute to another city using an airplane, that person generally thinks of flying from one airport to another. In reality, this is a less-than-complete view of the transportation system. Most people drive to the airport (or, e.g., take a bus or taxi) and rely on another means to transport them from the airport to their destination.

The trip from airport to airport is simply one step in the overall process and, for the most part, will often be referred to as a movement within (or using) one mode of transport. Although each mode of transport functions as a system, one cannot limit one's view to just one mode of transport at a time. This is largely due to the interdependencies between the various modal systems. It seems better to define the transportation system, in its pure sense, as being one layer above the various modal networks.[1] To go back to the airline flight example, the passenger leaving his or her home enters into the interrelated transportation system and relies on the various modal networks and their connections to move him or her through that system.

Modes of transportation generally describe a transportation system defined in terms of uniquely shared characteristics. These might include the technology associated with the maritime industry or rail industry. Generally, four modes of transportation exist:

(1) surface (or ground)
(2) rail
(3) marine
(4) air[2]

Pipelines can be another mode of transportation, as they view a transportation mechanism for moving fuel or raw goods in either compressed gaseous or liquid form.[3] Pipelines move a product within a stationary device, the pipe, whereas traditional transport mechanisms involve keeping the persons or goods on board or within a moving medium. For the purpose of this work and given the size of the work already, we will forego further detailed examination of this particular perspective, as this would complicate matters. This work will only regard transportation systems with the standard view of transporting people or things from one point to another.

2.2 General Overview

Transportation, like any other business effort, exists to meet a demand to move both goods and people while generating some kind of profit for those providing the system.[4] The only exception to this is when governments, using public funds, maintain a line of transportation to meet national needs. For those associated with securing the transportation system, one has to be careful to bear this demand (or need) in mind. This demand is a significant factor in the configuration of the transportation system, and, as such, it will have a direct relation to value, criticality, and impact associated with the personnel, assets, facilities, information, and activities that contribute to the achievement of the system's goals.

2.3 Nodes and Conduits

Nodes describe a pause in movement within transportation processes. A pause may be the result of a transition between modes (e.g., trans-shipping[5] a container from a ship to a rail car) or the demand to replenish supplies (e.g., fuel) due to technological limitations. Consider, for example, flight between two cities. The airport actually acts as a transition point between the surface modes (e.g., mass transit, cars) and the air mode. On the other hand, gas stations offer a service that allows for the continued movement down a conduit. In this case, the key differences are that transition points will involve the movement to a new mode, marking a terminus in one phase of the movement. The other mode will have no such transition but will have an interaction that should consider looking at elements of risk shifting in movement along the conduit.

One aspect of a node (as a transitional entity) is traffic or movement must funnel through a node to proceed further within the system. This creates a system choke point. At these points, the system is particularly vulnerable to disruptions as traffic on both sides of the nodes is disrupted.[6] In the previous flight scenario, loss of a runway at an airport is an event specific to the node, but loss of capacity will affect all traffic (aircraft) coming into the node because of loss of capability (e.g., delays).

Conduits describe the routes and connections between the nodes in the system. Generally, this will involve two parts: (1) the path taken to move between nodes; and (2) the means of moving across that conduit. These two components interact to determine the level of flexibility available in terms of time, geography, and capacity available for that movement. For example, an aircraft will be naturally limited in terms of its capacity to land and to take off, to carry up to certain weights, and to remain in the air for periods of time. On the other hand, trains may be limited in terms of the rail structure available between the nodes and considerations based on population centers, weight carried, and speed.

Conduits may be considered as *fixed* (e.g., roads, railways) or *flexible* (e.g., flight plans, shipping routes). The position and configuration of these routes will depend largely on a combination of meeting several factors, which include the following:

- Demand (for the service of transporting something)
- Regulation (e.g., environmental controls, pollution prevention)
- Geography (e.g., landscape, depth of water)
- Efficiency (e.g., great-circle and least-cost routes)

Fixed conduits, by their very nature, do not change. A rail line is static and predictable in terms of location. It provides only one avenue in which the conveyance can move. A rail line can make routing changes, but only at hubs or ports, as opposed to other modes (e.g., shipping, air transportation) that have abilities to make immediate changes to their route at any moment in time.

Care should be taken not to confuse networks of fixed routes with flexible conduits. Although the sheer number of permutations and combinations of routes offered may give a level of flexibility, they should still consider conduits as fixed in that they cannot change in response to its environment. A downtown core may offer several options when looking at moving between points, but people still need to use the road infrastructure to move between those points.

Flexible conduits are not as predictable since they can change depending on conditions in environment. Aircraft pilots may alter aircraft routes around significant storm systems to avoid turbulence where conditions threaten the safety of the crew and passengers. Similarly, a ship's captain may choose to chart a course around a storm if the storm appears to pose similar safety issues. Nodes may impose levels of control on conduits (e.g., initial approaches to an airport, landing patterns, anchorages) to maintain levels of order and predictability within any given system.

In some cases, fixed and flexible conduits may be intermingled as movement occurs between nodes. These conditions may exist where a normally flexible route is forced to move along a fixed route past a certain condition. These circumstances may involve the movement through passes or channels. These cases really represent a reduction in the flexibility of the route itself, but not to the point where a node is created.

2.4 Directly and Indirectly Derived Demands

Individuals using transportation systems create primary and secondary demands within the system. *Directly derived demands* involve primary movement of persons and goods between points. During the trip to the airport, taxi industries, mass transit systems, and airport limousines attempt to meet and compete for the demand to move people from their homes to the airport and vice versa.

In the background, these competing industries will use fuel, triggering a demand for the delivery of fuel to gas stations. Other parts of the industry that support the drive to the airport (e.g., auto parts, chemical) will require the transport industry to move goods from manufacturing point to point of sale. These represent demands that are *indirectly derived* from the demand to get to the airport. The services associated with meeting these indirect demands serve to support the direct demand for movement.

This level of demand also operates on the grand scale within the transportation system. Ports, airports, and train stations all exist to meet a demand to service the movement and trans-shipment of persons and goods through their respective facilities. In essence, the port and the ship exist in a reasonably symbiotic relationship where the capabilities of one align with the needs of the other. On the other hand, the presence of that activity spawns a number of secondary activities ranging from the movement of goods and services in support of the movement of the ships, trucks, and trains to the provision of basic creature comforts associated with either moving through or working at those facilities.

2.5 Factors Affecting Directly Derived Demands

Directly derived demands stem from the need for transportation and are directly linked to the performance of a task or function within an area.

The transportation sector supports these through the creation of nodes and conduits that allow for the right kind of movement between those with the demand and those able to meet that demand. This right kind of movement, identified by a specific mode of transportation, largely depends on operating environments and if there are substitutes (elasticity) that can replace the mode of transportation in that respect.[7] As a result, some may look at the following factors as being of primary interest when determining how far the direct demand for transportation actually exists within the system:

- Social or population: These stem from the demands of the population to have access between points. Consumers see this factor at play every day when looking at the industry as it shifts routes and creates new routes between communities to popular destinations. It is also exists in the movement of persons between urban centers. In these cases, the population center serves as one node while the desired destination serves as the other. The route in between will be determined based on the appropriate mode of transportation and the infrastructure necessary to support the safe and secure movement along that conduit.
- Infrastructure demands: These stem from requirements necessary to assure that a certain level of services are available within a given market or community. Certain services will be essential (or even critical), such as waste removal and, from a traffic control perspective, mass transit. The need to commute between living spaces and work creates a demand by population, but, in fact, demand actually is more a matter of infrastructure in how it promotes movement of people between a population center and its outlying communities. Nodes and conduits created by design to deliver services and, if possible, expand those services to meet the changes in need.
- Economic (supply chain): These stem from either movement of raw materials to areas where they are processed as part of the supply chain (e.g., coal mining and the rail industry or bulk cargo shipping industry) or the movement of processed goods from the processing centers to their eventual markets (e.g., containerized shipping). The location of raw materials identifies the edge of the transportation system, while nodes and conduits are aligned, and are based on convenience associated with collection points and their eventual destinations along the supply chain.
- Political: These stem from the need to ensure that infrastructure can be maintained in order to assure that sovereignty and similar issues are addressed. This generally involves a need to exert certain kinds of controls and the political will to expend those resources necessary to either establish or reinforce those controls. At the near end, the departure point of the resource sent to exert this control identifies one ode while the area controlled identifies the far node. Nodes associated with replenishment

and similar activities not only play a significant role in allowing movement along conduits but also provide a point from which flexibility can be built into the system.

These direct demands determine the outer boundary of the transportation system and can be limited through sociopolitical factors (e.g., conflict, cultural), economic (e.g., commerce, trade, profit), geographical (e.g., mountain ranges), and environmental (e.g., in terms of the ability to sustain viable settlements). Where no direct demands exist beyond that point, it becomes difficult to justify the needs to maintain business expenditures associated with maintaining links between nodes.

Consider another flight scenario between the eastern seaboard regions of the United States versus London, England. If there is a viable market to move people between those destinations, but not on to Europe, the outer limit of that system can be defined by the market served by London. This does not preclude that the outer limit of a viable service could overlap with another system (e.g., A flight from London to Manchuria may be viable for an airline, but not from the eastern seaboard region of the United States to Manchuria) such that a longer movement is accomplished. These limits generally reflect direct demands in markets that make those business activities viable. What should be clear to the reader is that where these overlaps exist, the size of the overall system grows to reflect the new exterior boundary.

Where the technology associated with the preferred mode of transport cannot meet the demands placed on it (e.g., range), a node is created that services needs of that technology so that movement may continue. In the case of aircraft, a refueling point may be created so that the aircraft has adequate fuel for the full movement. As long as the costs associated with meeting demand to move people between points of origin and destination do not exceed consumers' willingness to pay for service, service will likely remain viable from a business perspective. Where costs associated with overall movement exceed the consumers' desire to pay for service, direct demand collapses inward until the balance between the consumers' willingness to pay and industry's willingness to provide is restored. The viability of efforts to meet indirect demands is often vulnerable to shifts in these nodes. Much of this is based on principles of supply and demand economics; thus, demand varies depending on how much consumers are willing to pay for travel services (Figure 2.1).

2.6 Factors Affecting Indirect Demands

These demands stem from the presence of markets that require support for activities and, as a result, are predictable in terms of general market

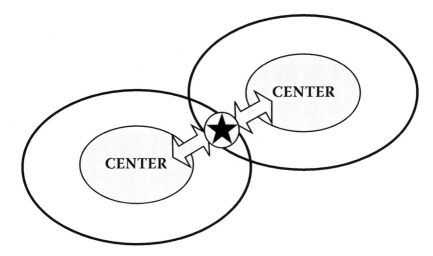

Figure 2.1 Nodes created to establish effective lines between centers.

pressures associated with supply and demand.[8] These generally include the following:

- Demands that are associated with the size of the population and the need, as well as its willingness to pay for goods or services
- Supply issues associated with availability of resources, costs of inputs across supply chains, and ability to maintain economic viabilities of enterprise profits

Demands associated with the size of the population relate to rates of consumption at specified points or within certain areas. In North America, it is common knowledge that in the early days, settlement and transportation routes were inexorably interlinked. These populations demanded a level of service due to their proximity to these routes, and, as populations grew, shipment along those routes also increased in terms of the volume and types of goods.

When conditions such as these exist, planners have two options. The first is to base the costs associated with transportation, particularly mass transit, in such a way that an acceptable level of profit is built into the base cost of each movement. In this model, the cost becomes part of the price passed on to the consumer and operates based on the principle that a reasonably predictable and assured number of movements (e.g., the necessity to bring the minimum supplies into an area) generate the necessary profits. On the other hand, the planner can use other factors, such as reducing the profit-margin-per-unit transaction, instead counting on an increase in the number of transactions to drive profitability. These two systems operate in fierce competition when looking at the indirectly derived demands.

In some cases, indirect demands will not be constant over time. This is particularly the case where technology (e.g., fuel consumption) demands inputs at a certain rate. Initially, the position of this kind of node is based on the need for technology to replenish itself, such as one might find with gas stations on a highway. As a result, the need to meet the direct demand actually fuels the system that creates the indirect demand along the conduit.

This may not be stable over time. As technology improves in terms of efficiency (less fuel per unit of distance), demand created due to the previous limitations gradually reduces itself. The result of these improvements in technology is a level of instability in the nodes previously established to support the original movement. Consider again the flight from the eastern seaboard region of the United States to London. If, for example, an aircraft could travel one half of the distance as opposed to one third of the distance, the natural tendency is to reduce the amount of potential waste associated with any additional stopping time.[9] Thus, the reduction in the sum of the support needed to accomplish movement creates competition between nodes in the system. This increased competition can result in a loss of viability of some of the infrastructure (Figure 2.2).

The positions of nodes that define the transportation system, in its macro sense, are determined by the direct demands present in the system. The supporting nodes are positioned in such a way that they best support direct demand. Thus, within the outer edge of the system (where direct demands exist), there may be a reasonable degree of stability as long as the demand to link to that node exists.

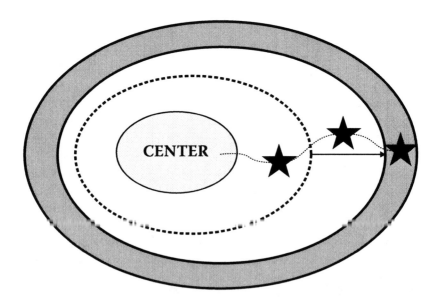

Figure 2.2 Nodes may need to shift or may shift with changes in technology.

Where the limitations of technology require there to be additional nodes, nodes will generally exist, in a competitive environment, to meet the needs of supporting the route. These nodes, however, will be less stable and will be largely dependent on the rate at which the technology using the conduit becomes more efficient and on the continued viability of the route. These nodes may also act as a downward pressure on their own viability as they require their costs covered to deliver the service. As these costs increase, the viability of the route itself decreases, and, for as long as the demand exists, those using the route will attempt to reduce those costs through several cost-reduction methods, such as contracting or improving efficiencies. Gradually, any node that is not delivering a valued service will be squeezed out of the system.

2.7 Routing of Conduits

As noted earlier, conduits can be either fixed or flexible, depending on the mode and environment. Conduits are generally arranged into networks that connect nodes. The arrangement of these networks will depend on business pressures (e.g., profitability, assurance of use, regulatory constraints) and environmental pressures (e.g., geography). Depending on markets and outer limits of direct demand, networks are established such that larger networks are configured until demand across the overall market has been reasonably satisfied. Looking at how these are configured, systems are organized to ensure that maximum capabilities of routes are maintained as best as possible (Table 2.1).

2.8 Spoke-and-Hub Systems

The spoke-and-hub system answers the question of how to maintain level of demand. Consider the point-to-point system that would exist if every city attempted to connect. The number of routes needed would be one less than the number of cities involved in the system. The problem here is that the system is vulnerable to weak routes (e.g., a route with low demand and high costs due to distance), and this weakness puts a downward pressure on the viability of the overall enterprise. The spoke-and-hub system uses central collection points such that goods and people are collected into central points, moving between

Table 2.1 Spoke-and-Hub System Example

Mode	Node Examples	Conduits Examples
Surface	Bus station, bus stops, distribution centers	Roads, bus routes
Rail	Train stations, rail yards	Tracks
Marine	Marine facilities, ports	Shipping routes, traffic routes
Air	Airports	Flight paths

them, then sent toward the outlying areas serviced by the central point. This stabilizes the demand for the route between the central points while reducing the business risk associated with the various smaller, outlying areas.[10]

2.9 Control Points versus Nodes

A *control point* is something that provides a service to coordinate movement along the node. While the node provides a service, the decision as to whether or not to stop at that node determines the entity in control of movement. As the driver of a car, that individual may decide that it is time to stop for services at a gas station; there are no laws or regulations forcing the driver to stop or to acquire gas at that particular station. A control point functions in such a way that it limits, or even prohibits, entities within the system from operating outside of what the system considers acceptable parameters. A traffic light or stop sign might be considered control points as they build a level of acceptable (or tolerable) predictability into the system by demanding that a driver stop before proceeding. The driver has no option but to obey the requirements associated with the control point since, should he or she disregard them, the driver is subject to a number of penalties associated with the severity of the violation as his or her decision created a level of unpredictability interpreted as risk within the system.

2.10 Control Points in Fixed Conduits

The goal of control points in fixed conduits is to build predictability into the system so that potential risks arising when conduits cross are reduced to acceptable levels. Two categories generally define these points:

1. The ability to move safely down the conduit based on a combination of design, engineering, and assumed capabilities (e.g., traffic calming circles or speed bumps), which are established to ensure that movement remains within designed specifications
2. Establishing a point where certain rules apply to control the interaction between conflicting interests (e.g., described in the case of the aforementioned traffic light).

Control points share a number of characteristics that include the following:

- They appear at locations where conditions exist in terms of geography or a location in space (intersections).
- They are generally fixed at those points, giving them a level of predictability (part of the infrastructure).

- They operate under set conditions within a system (traffic control systems).
- They are supported by some kind of oversight that is backed by legitimate force (e.g., traffic laws enforced by police).

2.11 Control Points along Flexible Conduits

Control points along flexible conduits do not necessarily share all of these characteristics, although they might share some depending on the specific circumstances. These tend to operate based on the following:

- They appear in situations where certain conditions exist in terms of the interaction between various components in the system.
- They are tied to situations in terms of predictability, meaning that they are predictable in light of the presence of conditions at any point in the system.
- They may or may not exist as a component within a system.
- They are supported by some kind of legal oversight (Table 2.2).

2.12 Terminal or Transfer?

Generally, when looking at modes of transportation the nodes are regarded, somewhat incompletely, as termini with respect to the movement of goods or people. This is reinforced by regulatory and other systems that tend to look from the top down, whereas organizations making decisions tend to focus on one issue.[11]

Most people, however, look at their total experience with the transportation system when planning a trip—let us say a flight for a business trip or vacation. To make a flight, a person must leave his or her home at a certain point in time, must have a certain amount of time at the terminal to pick up a boarding pass, and can then board the aircraft. Having arrived at the destination airport, however, the person would need to find a taxi (or rent a

Table 2.2   Control Points Example

Mode	Fixed Conduit Control Point Examples	Flexible Conduit Control Point Examples
Surface	Stop signs, traffic lights	Passing lanes, U-turns
Rail	Sidings, crossings	Rules regarding sidings
Marine	Navigation points, channel markers	Right of way
Air	Initial points, approaches	Near-miss rules

Table 2.3 Terminus Examples

Mode	Node Examples	Terminal and Transfer
Surface	Bus station, bus stops, distribution centers	Transfer within the surface and potentially into the mass transit system
Rail	Train stations, rail yards	Terminal within the rail system, transfer points between rails and other modes
Marine	Marine facilities, ports	Terminals within the marine industry, but transfer points into the rail and surface systems
Air	Airports	Terminals for air traffic in terms of a single flight, but transfer points for passengers into other flights or into modes, generally surface

vehicle) and then travel to his or her final destination for the remainder of the trip. While an individual may break the movement down into its logical components, he or she will generally not consider the movement complete until after actually arriving at the intended destination.

To examine security within the whole of the transportation system, the nodes must be thought of not in terms of a terminal[12] but in terms of a transition point where individuals move from one mode to another mode or into new conduits within the same mode.

Understanding the node in this context is crucial to applying security to the transportation system context. If we look at nodes as facilities or termini, we tend to view security in terms of simply protecting ends of an operational chain. Within the transportation system, security needs to take into account a variety of ongoing operations and throughputs into, within, and then past the node until the final destination (terminus) is reached. The risk apparent within the nodes, when looked at in this way, becomes a fluid entity that depends on the interaction between the stable environment inherent to the node and entities moving into the node (or even in proximity to) or out of the node. This will have profound impacts on the approach necessary to complete meaningful threat and risk assessments, which are discussed later in the work.

Table 2.3 describes some of the nodes within each of the modes and their relationship to other modes of travel. Looking at the security posture of the individual nodes, many will have to take into account the individual facility's operations, including its interactions with other modes.

2.13 System as a Sum of Interlinked Systems

As noted in Chapter 1, the transportation system is really a system composed of many other systems. Consider the commute to work again. This involves various conduits interacting with each other. When driving a car, most people generally cross points where roads (conduits) intersect—intersections. These conduits are all subject to a number of restrictions that allow them to interact predictably and that require persons to meet requirements of the control points described previously (e.g., a railroad crossing that intersects with a highway). The system relies on these control points, and the various levels of oversight associated with them, to ensure that the system is performing in a predictable manner. An individual that may decide not to stop at the various stop signs or red traffic signals (control points) runs several risks:

- The risk of those responsible for the oversight of the system removing them from the system (e.g., suspension of license)
- Forced change of behavior so that it falls within those acceptable parameters again (e.g., traffic enforcement, imprisonment)
- Injury or death from interacting with those who did observe the control point

This oversight is not limited to physical entities. Where control points integrate signals, the transportation system relies on a vast network of sensors, communication systems, and processing centers to maintain flow within the network. On the commute to work, many may pass through a number of stoplights. Incorporated in these systems are a number of controls (physical and procedural) that reduce the risk of two lights showing green in opposite directions. At a layer above this, these individual control points may be integrated into larger systems that are designed to improve the efficiency of movement within the system (discussed later in this chapter). Concurrently, procedural systems of oversight, including maintenance personnel, also watch for faults across these levels of control systems. The understanding of the interaction of these control systems, control points, and movement within the transportation system plays an important role in transportation system security to be able to prevent an attacker from disrupting the network. In this context, we can look at systems in terms of a collection of components organized to accomplish a specific function or set of functions (Table 2.4).[13]

Table 2.4 System Examples

Mode	Systems Examples	Example Indications of Subsystems
Surface	Bus, light rail	Spoke-and-hub systems, collection points, transfer points
Rail	Train stations, rail yards	Spur lines
Marine	Marine facilities, ports	Ports of call versus destinations International shipping versus short sea shipping or coastal trade
Air	Airports	Connecting flights

2.14 Recap of the System

Thus far, the physical layouts of the transportation system include the following:

- An outer limit defined by direct demand, which involves reasonably stable nodes
- Internal nodes along the conduits that support direct demand due to limits in, for example, technology
- Internal nodes that support indirect demands resulting from needs to maintain services meeting direct demand
- Conduits that provide the ability to move goods or passengers between points as efficiently as possible
- Modes using these conduits that depend on physical needs associated with conduits in terms of geography and time
- Nodes regarded as being transitional between modes of conveyance rather than as termini

Consider an example using the movement of bulk cargo. To send a shipment of coal to Japan from British Columbia, Canada, the coal is loaded using special equipment onto rail cars, which move by fixed routes to coal terminals at ports where the coal is then loaded into bulk cargo ships for shipping. On arrival in Japan, the coal is unloaded using special equipment into rail cars and is taken to its destination, where it is unloaded into storage facilities until it is used. This very basic example describes a number of nodes directly associated with movement of coal (the mine in British Columbia, a port in Canada, the Japanese port, and the rail unloading point) and a number of conduits, including both fixed rail and flexible shipping conduits.

Considering nodes to be a terminus tends to be a very narrow (if not skewed) view of the transportation system. The key is understanding how risk changes throughout the movement of the goods and cargo and seeing the various nodes as points where different risk systems intersect in such a way that an awareness of the change in risk can be comprehended.

A number of nodes involved within the process are not immediately apparent. These might include points that support the mechanics of the movement (e.g., fueling points) or the more human-needs driven elements of the movement (e.g., shore leave). These various elements are not necessarily subject to the same requirements. Thus, the potential of having systems maintaining themselves via disparate methods now exists: One involves the direct demand to move the goods between points; the others are mechanisms used to support that activity.

2.15 Constraints within the System

Given that much of the global supply chain functions based on a "just-in-time" system, the first obvious constraint is getting material to the location on time. This involves ensuring that steps within the movement of goods (or passengers) do not exceed anticipated or required dates (or even times) of arrival. For those who have had to catch connecting flights (particularly those with little time to spare), this constraint is more evident.

The second constraint is efficiency, particularly given the associated costs added by transportation into the cost of production and distribution. Efficiency will define whether specific processes or systems are contributing to success within the system by not adding unnecessary inflationary pressures on the costs associated with the movement.

Looking at efficiency and time, the perception begins to look at aspects of space. If it is neither necessary nor efficient for something to move, then the system will attempt to hold the item until it becomes either necessary or efficient. This means storing it somewhere, leading to geographical and physical constraints. Several conditions may influence this decision, including the cost of land, limits of weight that can be held safely by certain structures, and the availability of technology to handle the items being stored within the area.

2.16 Coordination Networks

The movement of people and goods is a coordinated activity that involves the ability to link and transfer between different parts of the same mode (e.g., transferring between buses in mass transit) and even between modes (e.g., taking the taxi to an airport to reach a flight).

This means that for each transaction, demands need to be communicated linking that demand to an appropriate service. This involves establishing levels of awareness with respect to what capability is operating where (time and space), what capacity it offers, and whether or not the client accepts that capacity (a booking) in as close to real time as possible. This process does not necessarily have to occur at the start of the process. Actually, it may be necessary during the movement depending on disruptions or other events that require a shift in the alignment of nodes and conduits.

2.17 Coordination Network—Operations

The client generally initiates this kind of coordination requirement. Consider the Air Canada on-line booking system. At the front end, the client selects his or her point of origin and then the destination using a standard interface, which serves as both the point of sale as well as the initial plan for the route.

Consider as an example booking a simple flight from Ottawa, Ontario, Canada (a national capital and transportation hub) to Prince Rupert, a developing transportation node. This interface is presented to the client and is available for any potential client who has access to the Internet. What this kind of information system enables the organization to accomplish can be interpreted from a number of factors:

- Client service: Since the client selects the flight, including its connection, there is a chance that customer complaints may be avoided.
- Predictability: At the first opportunity (the client's decision), the flight is booked and financial payment collected. This information can then be forwarded for inclusion in, for example, flight manifests.
- Supply chain efficiency: With selections (presented later or captured in the profile) for specialized items (e.g., vegetarian or diabetic meals), these can be incorporated into orders to arrive for certain flights at certain times, reducing the possibility of waste as opposed to attempting to constantly survey demographics to determine the basic load of an aircraft.
- Demand issues (directly derived): Systems such as this allow for routes to either be tailored to cover off potentially weaker routes (where a route cannot be deleted) or for a route to be identified quickly for discontinuation due to lack of demand.
- Demand issues (indirectly derived): By tracking potential loads for any given flight, issues such as ensuring that there are adequate supplies and stores for the aircraft become much more predictable. For example, if only 20 persons are booked on a flight with a capacity for 100, it might be assumed that the plane will not require the same amount of fuel (e.g., fewer regulatory constraints).

2.18 How the Coordination Network Interacts with the System

Supporting these factors just outlined is not unusual. These systems are such that information captured from clients is fed into trusted systems through protected channels. For example, the client may interface with a Web server that sends data into more trusted and critical systems to support the operations of the organization.

2.19 Conduit-Based Networks: Operations and Deployment

Whereas the administration of organizations relies on systems ensuring that assets and inputs meet at the right time and place, another series of systems requires that modes of conveyance use the right amount of inputs make their destination on time and in fair or good condition. These generally involve two kinds of systems:

1. The first set identifies resource amounts available to modes of conveyance. With a transatlantic flight, this might be fuel gauges on the aircraft at a very simplistic level but may also involve the reports to the airline regarding the amount of fuel or other resources available at stopover points. These kinds of systems indicate if the actual modes of conveyance are operating within acceptable parameters.
2. The second set deals with movements of modes of conveyance along a route. These kinds of systems might include navigational systems, altimeter, and other devices that inform crews that the aircraft proceeds along routes assumed to be safe and efficient.

The two kinds of systems often communicate with each other. For these cases, the information from one system is compared with the information from another, and a decision is made based on whether the results are within set or determined parameters.

2.20 Use of Systems for Automation

The use of automation is one trend that occurs frequently within the transportation sector. Automation, and the use of supervisory control and data acquisition (SCADA)[14] systems, appears to be increasing, mostly because machines can perform repetitive tasks more efficiently and in a more cost-effective manner than humans.[15] Where once beacons and relatively static infrastructure had been isolated, today even certain ports are looking to automation to increase the perceived efficiency of their systems.[16] Automation generally includes incorporation of inertial or global positioning systems (GPS), which are navigational systems that help allow bomb-carts and other equipment to move containers within the port. The autopilot is another automated tool that, through a series of sensors, maintains a safe operating mode of conveyance along a set path. The specific impacts, risks, safeguards, and vulnerabilities associated with automation are discussed later in this book.

2.21 Persons and Associations and Networks of Persons

Whereas laws and social agreements may limit relationships among clients, providers, employers, and workers, certain aspects of the transportation sector are reasonably conventional. For example, employees of a company may or may not belong to collective bargaining entities. Organizations need to be aware of emerging trends in employment, particularly self-employment and subcontracting.

2.21.1 Aviation (Air)

Many of the airport authorities directly employ personnel. This may include authority management and personnel within the industry. Certain roles are contracted to companies to provide services (e.g., Cara Operations Limited for providing airline food), and, in some cases, key roles are still maintained by governments for airport security (Free Trade Area of the Americas [FTAA][18] in the United States and Canadian Air Transportation Security Authority [CATSA][19] in Canada). Key roles requiring specialized services may involve contractors or self-employed individuals.

2.21.2 Marine

A special arrangement exists between many ports and their labor forces when marine facilities hire people. The engagement of personnel is handled through employers' associations that deal with the local collective bargaining units, such as labor unions or trade guilds.[20] This creates pools of available labor that are reasonably flexible to match rising and falling demands, but also lends a level of unpredictability in terms of who will be working at the facility at a specific time. Similarly, companies or subcontractors may be hired to perform nonunionized tasks.

2.21.3 Rail

Within this industry, many of the employees directly work for the company, with collective bargaining entities representing groups of employees. While some tasks may still be subcontracted (e.g., specialized services), this mode is generally very stable in terms of its workforce, arguably the most stable.

2.21.4 Trucking

At the other end of the spectrum, the trucking industry has significant representation of company drivers, some with representation, and there are self-employed or freelance drivers.

2.22 Sector-Wide

Recognizing that the transportation system, in its sector context, is transnational in nature is another important step when looking at securing this infrastructure and it capability. It leads to a level of fluidity. This fluidity reflects a need for the transportation system to be able to meet its direct demands to accomplish movement. Although this movement tends to take place through nodes and conduits, it must be understood that these may be subject to several, and even disparate, structures that have tended to align themselves on administrative lines. For those involved in attempting to secure this system and its capacity in a meaningful way, understanding the shifts between systems and the role of nodes as transition points within this system will be crucial.

2.23 Factors to Consider

You have decided to take an overseas vacation with your family. How will direct and indirect demands will influence your decision?

- Depending on the time available for your vacation and the distances and other factors involved, you will likely select a mode of transportation that fits both your budget and ability to meet those needs. The direct demand is your desire to go to the vacation point. The indirect demands will play a significant role in the costs associated with your decision.

In your vacation plans, what modes and conduits will you use? Are they flexible or inflexible? How do they act as transition points?

- In taking the taxi to the airport (assuming that the air mode provided the best option), you will use a network of fixed conduits that allows you to get there on time. Having arrived at the airport, you change from the surface system to the air system (transition point). During the flight, which follows a relatively set route, the pilot is able to reroute around several storms (flexible route) until reaching the initial point for the approach to the airport. Having arrived at the airport, you will then switch back to the surface mode until you reach your vacation point (assuming you are not taking, for example, a cruise or train).

Notes

1. When looking at supply chain management, for example, articles such as those put out by *CFO Magazine* have identified the linkage between ports and other modes of transports as being a major challenge in supply chain risk management activities. John Goff, "Delayed in the USA," *CFO Magazine,* http://www.cfo.com/printable/article.cfm/7851829?f=related.

2. "About Critical Infrastructure," Public Safety Canada, http://www.psepc-sppcc.gc.ca/prg/em/nciap/about-en.asp#1; "Homeland Security Presidential Directive/HSPD-7," http://www.fas.org/irp/offdocs/nspd/hspd-7.html.

3. Many systems include pipelines in the transportation sector, and their characteristics are comparable.

4. Paul M. Johnson, "Demand," *A Glossary of Political Economy Terms,* http://www.auburn.edu/~johnspm/gloss/demand.

5. The term *trans-shipping* is in reference to a form of transcontinental shipping, often associated with modal or containerized freight and cargo.

6. John J. Jarvis and Duane D. Miller, "Maximal Funnel-Node Flow in an Undirected Network," *Operations Research: Mathematical Programming and Its Applications,* Vol. 21, no. 1 (January–February 1973), pp. 365–369.

7. A. Marshall, *Principles of Economics,* 9th ed. (London: MacMillan, 1961), Volume 1, Book 5, pp. 381–393.

8. ibid., Book III, Chapter 3, http://www.eco.utexas.edu/~hmcleave/368 marshallonwantstable.pdf.

9. "Gander International Airport," *Wikipedia,* http://en.wikipedia.org/wiki/Gander_International_Airport.

10. Concept derived from J. J. Coyle, E. J. Bardi, and R. A. Novack, *Transportation,* 4th ed. (New York: West Publishing Company, 1994).

11. Consider, for example, the International Ship and Port Facility Security (ISPS) Code put forward by the International Maritime Organization (IMO). The ISPS code itself speaks very specifically with respect to ships, vessels, ports, marine facilities, and other entities but does not mention how to handle the interaction with the landside of the transportation system beyond the interaction of the ship-marine facility.

12. In the Latin vocabulary, the definition is called a *terminus,* which means "ending point"; "Terminus," *The Free Dictionary by Farlex,* http://www.thefreedictionary.com/terminus.

13. Interoperability Clearinghouse Glossary of Terms, *Interoperability Clearinghouse* http://standards.ree.org/reading/ieee/sto_public/description/se/610.92_1990_desc.html.

14. "Critical Infrastructure: Control Systems and the Terrorist Threat" (RL31534) (Updated February 21, 2003, Dana A. Shea, Resources, Science, and Industry Division, Congressional Research Service, Library of Congress), http://www.fas.org/irp/crs/RL31534.pdf.

15. Ibid.

16. Behrokh Khoshnevis and Ardavan Asef-Vaziri, "3D Virtual and Physical Simulation of Automated Container Terminal and Analysis of Impact on In Land Transportation," University of Southern California, December 2000, http://www.metrans.org/research/final/99-14_Final.pdf.

17. http://www.cara.com.
18. The FTAA is an attempt to expand the North American Free Trade Agreement (NAFTA) to every country in Central America, South America, and the Caribbean, except Cuba; "Free Trade Area of the Americas," *Global Exchange,* http://www.globalexchange.org/campaigns/ftaa; official Web site, http://www.ftaa-alca.org.
19. http://www.catsa-acsta.gc.ca/english.
20. The specifics of the employer–employee relationship can generally be found through the Web sites for port authorities or the various employers' associations in Canada.

Business Goals and Mission Analysis 3

Objective: At the end of this chapter, the reader will be able to do the following:

- Identify the general mission for transportation systems security
- Identify tactical, operational and strategic levels and their application
- Differentiate between these issues at their respective levels

3.1 Introduction

Although the transportation system provides a complex and far-reaching series of nodes and conduits organized into networks, functioning as a system it incorporates two other concepts: (1) the transportation sector's interdependencies on other sectors; and (2) the strategic, operational, and tactical levels that must be examined when addressing overall goals and mission analysis of this sector.

3.2 Scales of Operability

Looking at the levels of operability within the transportation system, a special aspect of thinking has to enter into the bigger picture. Imagine yourself attached to a huge elevator. At sea level, you are within a facility and can see in fine detail the operations that are immediately surrounding you. The movements of personnel, assets, and so on are available to you clearly, but many cannot be seen much beyond the immediate area. As you progress upward, you begin to lose the detail of those operations but begin to see how other pieces of the facility interact with it. At 15,000 feet, you essentially lose any meaningful detail about what specific people are doing at that facility, but you can identify the ship, air, road, and rail connections, including the facility's clients and supply and support chains. Finally, as you move past 30,000 feet you lose this level of detail but can still see general connections between population centers in terms of major roads, railways and seaways that interlink major centers together. Finally, at 50,000 feet you can see the major population centers and how they are interconnected, while from orbit you can identify the major global centers on that side of the planet and can draw the lines between them.

It should be noted here that one challenge that has posed some issues between the authors comes from two distinctive uses of terminology describing management levels of doctrine. Certain systems tend to reverse, misuse, or interchangeably use the concepts of *operational* versus *tactical* management levels. From time to time, those with military experience (i.e., those who see the tactical level as the level closest to specific activities at a singular instance) will encounter those without military experience who term this as an *operational level*. For the context of this work, we look at using the military level of doctrine of *tactical* as the lowest denominational level and *operational* as one level higher.

In this concept, the view of the facility level is referred to as the tactical level, the level at which an individual's actions, training, and decision making have an impact within that immediate organization locally. As one moves higher and loses sight of that level, one moves into the operational level. Here individual decisions still have an impact, but it is within a larger network or system of events. People who have a view of the operational level have a view of the network and system and make decisions across that scope. Within the business context, these represent terms of regional or district offices, organizations that are responsible for the performance of a collection of entities within a homogeneous administrative area. Finally, the strategic level is compared to the corporate headquarters, representing the strategic thinking of an executive. Moving yet higher, the incorporation of additional heterogeneous administrative areas forces the organization to identify common grounds and threads linking those organizations. This leads to a series of approaches, philosophies, visions, and goals set then passed down to regional levels.

Another perspective might be that at a tactical level, continuity of the operations is maintained but not modified. Personnel usually operate with a set of instructions given to them and are usually told not to deviate from these instructions. At the operational level, individuals' actions change the nature of operations (in a procedural sense) as the focus is to maintaining levels of performance. Here, personnel may make changes but will have to follow instructions, though this will involve persons' meeting goals and intents while being given some leeway with respect to specific actions or instructions. Both operational and tactical levels deal with present or current business functions. At the strategic level, future continuity of operations is determined based on environmental, market, or industry trends and analysis performed against competition or laws recently implemented that affect both the organization and the industry. There are no instructions because the strategic level sets the direction and pace at which changes will be implemented.

3.3 General Interaction

These levels of operability flow into each other and simply cannot be considered in isolation. Consider an analogy of the human body when it becomes infected. The overall goal of the human body is to remain alive and healthy (strategic aim or intent). To continue doing this, the human body follows certain rules regarding what it can do to remain healthy—for example, remaining clean, stabilizing body temperature, preventing the consumption of spoiled food or harmful things. The specific rules will change depending on the environment in which the body finds itself (operational). For example, the rules may be very different when looking at warm climates as opposed to colder climates. Similarly, the body must also be able to identify whether to eat food that is good or harmful (tactical). The key here is to ensure a level of cooperation and coordination that the overall system has the best possible chance of reaching its strategic aim.

Moving back to the transportation system, similar principles are at play. Consider the following questions:

- When security guards perform a task (or set of tasks), are they following steps applicable to their environment and supporting organizational priorities and goals?
- When a facility implements a security measure, does it take into account regional factors, identifying how these support organizational goals, and does it take into account the ability for personnel to deliver measurable results?
- When the organization develops a policy, does it take into account feedback from the operational and tactical levels to identify potential areas of weakness or improvement? In extreme cases, this may even include identifying that the intent of the specific policy cannot be achieved or realized.
- Does the organizational policy leave adequate flexibility for operational and tactical levels to work within their environment but also to identify those factors back to the strategic level?

In this context, the first key is to be able to ensure a unity of focus or command within the organization so that various subcomponents are not misallocating

Table 3.1   Issues in Various Levels of Management

Issue	Strategic	Operational	Tactical
Regulations prohibit the movement of dangerous goods unless certain conditions are met.	The company will only move dangerous goods when it can be shown that all legal requirements to move those goods have been met.	To meet the company policy regarding the movement of dangerous goods, the shipper must include the following within the documentation provided.	To meet the operational level, persons at the gate will collect and verify the appropriateness of documentation provided.
Provide feedback on how to improve abilities to meet this requirement	To date, the organization is meeting or not meeting this requirement.	Within this regional administration, we have had the following number of events of this a certain kind per number of shipments.	The event occurred because we are still being trained on the requirements. We could do this better if …

resources or competing with each other. Consider Table 3.1, which describes a relatively sound alignment of the effort.

3.4 How Is the System Mission Achieved?

This is one of the most challenging questions when looking at this issue from a strategic level. In business, corporate headquarters would decide the mission statement, vision statement, and similar goals. These intents would drive activities of the organization across all levels. So who is responsible for the transportation system as opposed to an organization that either uses or contributes to the system?

Currently, the transportation system is in a state of evolution. Oversight operates on a mode-by-mode basis. For example, the International Maritime Organization (IMO)[1] provides the central coordination for the maritime industry while the International Civil Aviation Organization (ICAO)[2] provides the same coordination capability for the civil aviation industry. These international organizations provide a focal point and generate conventions (e.g., the Safety of Life at Sea [SOLAS][3] overseeing marine safety) that provide focus to efforts for that mode and that each nation enshrine within its own legislation. The challenge is that the transportation system respects neither of these considerations. Why? The overall system is transnational in nature (with exception of mass transit at the local levels) and operates across several modes over vast geographic areas.

Each nation that signs on to these conventions generally has a lead organization with which the international community can maintain contact and that provides the source of authority to make appropriate decisions within the context of the convention and the link to the transnational nature of the system. These lead authorities, two of which are the U.S. Department of Transportation[4] and Transport Canada,[5] are responsible for generating the primary level of constraints of the system within their own borders.

3.5 Considerations of the Transportation System

The role of the transportation system primarily focuses on moving people or objects, with the points of origin and destination determined by the client. The routes taken may or may not be subject to controls. For example, a client may have no interest as to how the shipment arrives—only that it arrives at the destination on time and undamaged. Other kinds of shipments may be subjected to significant controls, such as those placed on the movement of a dangerous-goods list or items that are subject to export control lists.[6] Similarly, the sector has time constraints factored into it as well as quality assurance standards associated with the safety of the persons or items moved or transported. These constraints are arranged by contractual arrangements, service-level agreements, or other similar legal obligatory mechanisms.

There is another aspect to security within the transportation sector: the protection of infrastructure itself. This falls into two major categories:

1. The direct protection of the infrastructure against attack (e.g., protection against the use of explosive devices on aircrafts), which is in direct response to the perceived threat against passengers. It has been widely accepted that the means, opportunity, and intent to attack certain elements of the transportation system exists today and will likely exist well into the future.
2. The use of the transportation system as a conduit or the means to move something harmful as part of an attack. In this respect, the transportation system is not itself the target but provides an opportunity for an attacker to gain access to a target of value. For example, an explosive device may be moved through the transportation system to its final destination.

It should be noted that preventive measures are not part of the usual conduct of business for most organizations but only happen to legitimately conduct business. One of the great challenges within the transportation sector is to change management's view that preventive measures can act as a brake only with respect to performance to that of seeing preventive measures as a benefit. For the security professional, this is both challenging in terms of having to exercise levels of diplomacy (i.e., building arguments and putting them forward) and creativity (i.e., problem solving) when putting forward options to management. Given that the nature of the transportation sector involves using throughput as a significant measurement tool and that the transportation system's infrastructure plays a key role in achieving that level of performance, this cultural view of security as a brake may well be one of the most significant challenges to overcome.

3.6 System-Level Mission Statement

No single organization or entity has the necessary authority to produce a mission statement for the transportation system. It simply does not exist. Given what the transportation sector's role involves, some might define a relatively universal mission statement for the transportation system in terms of accomplishing the following goal:

> The movement of appropriately authorized goods and personnel between desired points, within an acceptable timeframe, such that it ensures that those persons or goods arrive at their destination within an acceptable condition.

This mission statement identifies reasonably clear objectives that can be shown using two examples:

1. The passenger industry, which moves authorized persons (i.e., passengers not known to pose a threat to the security of aircraft, passengers or the nation) between airports established to handle those aircraft safely within predictable schedules and in comfort
2. The cargo industry, which moves goods between points that are arranged with the client and ensures their timely delivery and arrival in acceptable condition

How does security operate at the other levels?

3.7 Transportation System Security Mission Statement

The security of the transportation system works on two fronts when considering the mission statement.

On one hand, the requirement to maintain performance within the system for such activities as trade, commerce, and movement of people requires protection of personnel, assets, facilities, information, and activities associated with delivering that service. In this case, there is considerable pressure to minimize the number of checks and balances that could possible interfere with the process and create wasted time in the system. For example, if the security checkpoints at airports offered little to speed the clearance process, lines would exists as long (if not longer) than those that initially existed shortly following the 9/11 incident in the United States; security screening verifications checks would consume two (if not more) hours of the passengers' time while they waited in line. The pressure to streamline the security screening checks represents such pressure to improve the efficiency of the system.

On the other hand, the concept of appropriate use[7] generally flows from the requirements or constraints imposed by national, provincial or state, territorial, and local governments in terms of what is acceptable within general communities and how they interact. It also encapsulates a level of management authorization to use the company's or entity's resources to achieve certain goals in support of its business lines. This may include a range of considerations from antiterrorism to crime prevention. For example, it is generally considered inappropriate (illegal) to smuggle goods across borders where there is a requirement to report those goods through the relevant customs organization; at the same time, it may be considered inappropriate to use fuel and other company resources for personal gain. Consider the mission statement for the transportation system security organizations to reflect something similar to the following:

> The transportation system security effort focuses on the maintenance of the transportation sector's ability to meet its mission through the assurance of continuity of an infrastructure necessary to meet performance goals while preventing the system's use by those who would use it for inappropriate or nefarious intent or purpose.

3.8 Determining the Mission Statement for Organizations

The mission statement for any enterprise answers the question: What do we do (to make money)? When these services are delivered in the public sector, the questions rephrase the reasons public funds are provided to the specific organization. This information can be found within business cases and similar documentation that defines the role of the company, the goods or services it intends to provide, and for what it is willing to commit resources. The transportation system generally involves a number of private and public sector entities, and each of these entities focuses on its own particular role, not that of the system as a whole.

It is important for security organizations to recognize this point and to be able to relate the specific protective measures to requirements beneficial to both the individual organization and to the transportation system. Relating the measures to the role of the organization and defining them in terms of the protection of personnel, assets, facilities, information, and activities will play a significant role in achieving the necessary level of acceptance from managers to commit scarce resources when faced with a number of organizational priorities. The failure to recognize this point leaves the system at risk as the core service that the system relies on may not be subject to an adequate level of protection. In this context, the security officer may wish to keep in mind that the failure of any part or relationship between parts within that system has an impact on the performance of the overall system.

It must also be noted that while the mission statement may reflect a security consideration, unless the organization provides a security service as its primary source of income it might not appear in the mission statement for the organization. This is because security, when functioning appropriately, supports the overall business efforts and does not function as an independent organization (even though many of its services may operate at arms' length to maintain a necessary level of impartiality). Most organizations consider security a liability to the overall financial success and profitability of the organization and therefore restrict its application. In short, security risk must be incorporated into enterprise risk:

> Example: ABC Transport will provide high-quality door-to-door services for the movement of special delivery parcels in North America. (ABC Transport is used throughout this book to provide similar experiences at different stages. It is a fictitious company that (hopefully) identifies issues, risks, and threats similarly encountered by this sector without identifying any one specific organization or industry.)

In this statement, several aspects regarding security are implied. For example, sending the parcels by special delivery suggests an additional level of value to the delivery service. As a result, there is at least a perception that certain demonstrable steps will be taken to ensure that the delivery will be to the client's satisfaction.

3.9 Strategic Level Mission Statements as Organizational Constraints

With the strategic mission statement determined, it must be understood that this mission statement provides the second level of constraint within the organization. The first constraint is the imposition of legal requirements on the organization from international agreements through the passage of laws, translated into various levels of regulation. The second constraint deals with what the facility considers an appropriate use of resources and focus of effort. Consider the mission statement of a Canadian rail company, which states that it "… will work together to provide travel experiences that anticipate the needs and exceed the expectations of our customers." (http://www.viarail.ca/pdf/an2006/viarail_ar2006_en.pdf) When this statement is deconstructed, the following aspects become more visible:

- "Work together" indicates a kind of relationship to be maintained both internally and externally.
- "Provide travel experiences" indicates that passengers must be able to enter into the system and receive a predictable service in return for remuneration. The organization commits to making this possible.
- "Anticipate the needs" indicates maintaining the ability to communicate with clients and determining the trends that would point to future needs within the industry.
- "Expectations" includes high service standards in terms of quality and arrivals at the right place at the right time, on time, and safely.

The organization's strategic mission statement provides the framework within which the various subcomponents of the organization must function to support the organization's goals. From the strategic level, the mission statement is refined to a series of requirements that can be implemented at the operational and tactical levels. The goals determine methods by which processes and documented procedures are established, written, implemented, and maintained.

If the mission statement at the strategic level is to meet the needs and expectations of clients that require shipping services, then the strategic level will communicate its own tools and expectations across all aspects of the organization to meet this need. Strategic-level mission statements may include the following:

- Undertaking any activity, the organization will seek out expectations of the client (constraint on performance).
- After identifying client's expectations, the organization communicates this expectation using the measurement system provided. This ensures that the organization is meeting expectations of the client. The measurement must be auditable and therefore verifiable by outside auditing organizations (quality assurance).
- The organization will commit resources (based on value of the client) to ensure that expectations are being met (constraint on operations and delegation of authority to allocate resources).
- When expectations cannot be met, the value of disruption to the client will be determined, and the client may be offered regular services in lieu of to the value of the loss (delegation of authority and customer service).

3.10 Operational Level within the Structure

The sum of the collection of operational levels makes up the strategic level. Each has a vertical interaction (with the strategic level), a lateral interaction (with neighboring operational levels), and an interaction as a controlling level (with the tactical level). For the overall mission statement to be realized by the organization, these interactions must be as graceful as possible, flowing from one into the other in such a way as to provide necessary levels of clarity of intent and authority.

3.11 Interaction between the Strategic and Operational Levels

Work within this structure cannot be performed in isolation—it must involve consultation between the levels. The strategic level, by identifying the impacts that need to be reduced (for a variety of sociopolitical perspectives), may be able to identify the system-wide impact to the overall system (note that the system has all modes, not just one).

To establish this framework between the strategic level and the operational level, the strategic level must do the following:

- Set organization-wide policies
- Establish organization-wide measurement systems
- Establish specific categories of information to be collected and sent to the strategic level
- Set specific goals of the organization, including objectives to be met by the tactical level
- State any aspects of residual risk that the strategic level will not accept in the system
- Requite baseline levels of performance or compliance with system or organizational requirements that cannot be breached

In return, and as a condition of resources and maintaining a delegation of authority, the strategic level demands that the operational level do the following:

- Understand that provision of resources and support is contingent upon acceptance of strategic-level constraints
- Maintain use of system and measurement tools integrating with the strategic level and not use tools that cannot be related to the organization's systems

- Return, as expeditiously possible, requests for information passed to the operational level
- Indicate if the operational level is meeting goals set by the strategic level as well as terms of lessons learned (where goals are not met) and best business practices (where goals have been exceeded) such that the system begins to learn and evolve at the strategic level
- Indicate whether or not the level of residual risk assumed by the strategic level is, in fact, valid and monitor any changes in those areas of risk, including the appearance of potentially unforeseen risks
- Report any situations where baselines or similar thresholds had to be crossed and what actions were taken to reestablish those baselines

3.12 Role of the Operational Level

The role of the operational level can be divided into three parts. While many organizations define which activities are more important and the corporate culture may swing on which function dominates at the time, it is important to realize that each function has its place in ensuring the success of the organization.

First, there is the maintenance of the strategic level's intent and goals. This means that the tactical level of the organization complies with the strategic level's intents and requirements. When the tactical level is not operating within acceptable limits, the operational level provides a layer of control to bring that entity back into line.

Second, the operational level works to allocate resources to ensure performance of the system during periods of planned or unplanned disruption. The operational level manipulates the system within its area of responsibility to minimize impacts while also restoring the system to acceptable levels of performance as much as possible.

Third, the operational level ensures that information requirements pass to strategic levels so that the system maintains both a level of awareness and the capability to learn. This might include conducting internal reviews, investigations, root-cause analyses, or other activities resulting from processes that generate lessons learned or best business practices at the tactical level.

3.13 Tactical Level within the Structure

The tactical level is the front-line client-services role of the organization. It delivers the service within the context of mission statements and focuses on the delivery of the expectations of the operational level integrating with the strategic level.

The tactical level is made up of a number of these entities, each providing a specific good, service, or function to the system. The interaction of these entities determines the success of the tactical level and can even have a significant impact on the performance to the overall system.

3.14 Interaction between the Operational and Tactical Levels

The tactical level has the responsibility of maintaining the performance of the system within its area of responsibility and operating within strategically derived constraints. These constraints may be refined by the operational level to modify factors (e.g., political, economic, sociological, technological, geographic) that can have an impact on abilities to deliver services, even before the operational level communicates them to the tactical level. For those working at the tactical level, these constraints become the defined procedures and expectations placed on specific roles or positions. They may include legally binding agreements between employer and employee or may include procedural documentation that has been officially endorsed by management who are accountable for performance of work related to those procedures.[8]

The operational level reports its abilities to deliver expected levels of performance and identifies specific successes or shortfalls and responses to the strategic level's management. This one component provides basic building blocks for strategic level awareness through abilities of the tactical level to indicate specific capacities at specific points within a system. The sum of this information, collated and assessed at the operational level, defines capacities within the operational layer of the organization that, when collated into an overview of the entire system, defines the overall capacity of the system.

Concurrently, the operational level is the first level of system awareness within the system and for those reasons plays a key role in responding to situations and generating lessons learned. The tactical level will provide the immediate response to events of interest (e.g., security breaches) with the means to contain the event and return to normal operations as quickly as possible. Part of this process involves the passage of key information that helps the operational level determine the impact of the event within the operational layers' context of performance of a regionalized system. Having contained the event, the tactical level will provide much of the work in the response and recovery phases. Finally, the tactical level must be able to communicate with the operational level in such a way that future events of a similar nature are avoided with best business practices and lessons learned.

3.15 Overview of the Structure

Although some organizations may place greater emphasis on the various levels of the organization (e.g., strategic brain trust, tactical-level grunts), this is not indicative of a smoothly functioning system. A functioning system recognizes that, as with any decision of significance or complexity, the key is to involve the right people with the right kind of information at the right place in the process so that the best result occurs. For managers familiar with the concept of total quality management systems, this approach should resonate in harmony with the concept that the whole organization is involved in ensuring the quality of the goods or services delivered by the organization.

Similarly, the involvement of the whole organization involves establishing a cycle to accomplish a system including continuous learning and adaptability. As the strategic direction is communicated to the tactical level, the tactical level provides feedback to the strategic level (both through the operational level) that essentially allows the senior management of the organization to determine if the core goals are being met (Table 3.2).

3.16 Limitations on Controls

As noted already, this kind of effort requires that the right kind of information is inserted into the decision-making process at the right time. Involving multiple layers of participation tends to be challenging when various levels of the organization begin to intrude on each other's roles and responsibilities, bringing chaos or friction into the system as the decisions they render are applied outside of an appropriate context. To prevent this, there must be adequate acceptance from all levels of management within the organization, but particularly the authoritative level of management. This ensures that the system's strategic, operational, and tactical layers are working harmoniously together in such a way as to avoid friction within the organization.

3.17 Limitations on the Strategic Level

The role of the strategic level must be refined by limiting its powers in terms of scope. With the focus of the strategic level being on the overall performance of the entire system, is it appropriate for that level to attempt to generate specific measures to be applied at the tactical level? On one hand, the strategic level has an obligation to refine and define its requirements in such a way that they can be used across the entire system with confidence that the process is capturing the right kinds of results. On the other hand,

Table 3.2

Level	Mission	Delegation	Key Activities
Strategic	Sets the organization's mission, working within certain constraints (e.g., national and international law)	Delegation in the private sector stems from key authorities in the organization	The setting of the mission, identification of specific goals, and then monitoring of the system in terms of results comparable to the goals being set
Operational	Works within the organization's mission, reflecting more localized constraints (e.g., provincial, state, territorial law)	Delegation stems from the strategic level, ensuring that lines of command and communication are clear and without conflict	Monitoring of the performance within the tactical area and making adjustments to maintain that performance by providing direction to the operational level. Tactical-level goals are developed based on strategic-level information requirements and are refined to specific tactical-level information requests.
Tactical	Works within the tactical level's constraints, respecting local constraints (e.g., city bylaws, tribal laws)	Delegation stems from the strategic level through the operational filter. The strategic level provides the source of authority whereas the decision as to specifics is incorporated within tactical-level delegations.	Development of specific measures or steps within the operational level's constraints and reporting the success of those measures in comparison with goals communicated by the tactical level and based on strategic-level information requirements

the strategic level may lack knowledge of key personnel, assets, facilities, activities, and information specific to various locations in the organization. This makes such a goal difficult and perhaps wasteful.

At the same time, the temptation for the strategic level to become involved in the tactical level must also be limited from the perspective of maintaining the overall system. It cannot simply abandon the other operational areas of responsibility due to a potential situation at the tactical level in one location. Doing so creates conditions in which services promised by the strategic level are not delivered. To manage the entire organization requires a level of discipline and trust in the tactical and operational levels of management.

3.18 Limitations on the Operational Level

The operational level focuses on a different aspect of the system and may face the most significant challenge in terms of operating within its scope. On one hand, the operational level must be able to communicate with the strategic level to convey information but should probably resist the temptation to make strategic-level decisions.

Still, within the operational level, monitoring the application of the strategic-level program may become deeply involved in attempting to resolve situations at the tactical level. This reflects a direction of an operational-level entity wishing to take a specific measure. Care should be taken at the operational level as operating without a full understanding of the situation on the ground can prove difficult.

3.19 Limitations on the Tactical Level

The main source of temptation at the tactical level involves the pressure to do what is needed instead of what is required while also slowly detaching from the apparently administrative requirements of the operational and strategic contexts. The first area of concern is the fact that the tactical level does not have the full information regarding the potential impact of its actions, particularly from the operational and strategic levels, and that making this kind of decision can open the organization up to unanticipated avenues of risk. Second, this may begin to deconstruct the organization, leading to a situation in which individuals are potentially operating outside of their delegated authorities.

For example, pressures associated with ensuring the security of the node during busy periods of operation may lead to periods of time when routine reporting requirements are not being or cannot be met. This could be the result of an undisciplined tactical level deciding that the reporting requirement puts undue pressure on the local facility, or it may also be the result of a strategic-level decision to place unreasonable demands in terms of information passage, such as may be common in cases of managers who micromanage their organizations. This is a particularly difficult situation, in itself, as it slowly evolves into gaps in the understanding of the system at the tactical and strategic levels that can seriously erode any ability to make timely and effective decisions across the organization.

3.20 Generation of the Mission Statements

Given the information presented herein, the following process may assist in the generation of mission statements across the strategic, tactical, and operational levels. Although this methodology is provided as one possible course of action, it is not the only solution, and, in fact, many larger organizations may well have components of their organizations that can greatly assist in the formulation, documentation, and communication of these kinds of messages:

- Define the overall mission in terms of what the organization wants to accomplish as an enterprise.
- Determine what support the organization's overall management will require so that appropriate and timely decisions made in support of organizational activities.
- Determine what support will be required from the tactical level identifies communications, and support lines necessary to bring that information to the strategic level.
- Validate the operational- and tactical-level information to ensure that the organizational focus is on meeting its overall goals.

The second possible course of action of this activity is determine what support methods will be required by tactical and operational levels and to delegate authority to deal with those situations as long as none of the actions taken comes into conflict with the strategic level. This involves both the delegation of decision making as well as of resources to put those decisions into play. Of course, the strategic level will likely place constraints on these decisions, requiring an escalation of the decision to spend funds exceeding certain thresholds and so on.

Once these two steps have taken place, the next step is to generate the mission statements in such a way that each statement clearly identifies and defines the role and intent of the role within the organization, linking it to the strategic-level mission statement.

Consider, for example, the following three statements, referring to the ABC example:

- ABC Transport will provide top-tier services focusing on the point-to-point, on-time, on-demand movement of goods for markets within North America.
- ABC (East Coast office) will provide top-tier services identifying potential markets and coordinating the movement of goods within the Eastern Coast Administrative Region in support of ABC Transport's mission.

- ABC (Halifax, Nova Scotia, office) will provide top-tier services identi-fying potential clients and coordinating collection, storage, movement, and tracking of goods from and to Halifax within the ABC Transport organization.

3.21 ABC Transport's Security Mission Statements

Remembering that the security organization provides two roles, the first involves the assurance of infrastructure required to meet certain key mission requirements. The second deals with ensuring that regulatory requirements are met so that the organization does not become perceived as a potential conduit for attack against the state it operates in or its citizenry.

- ABC Transport corporate security will provide the policy framework for security activities, will monitor the overall application of the security measures, and will identify areas of impact across the entire system or risk across the system in support of ABC Transport.
- ABC East Coast office security will provide guidance and advice, will gain approval for appropriate security standards and procedures in the effort to ensure the performance of the East Coast office's functions, and will provide direct support to the management in controlling impact and risk within this administrative organization.
- ABC Halifax security officer provides guidance and advice, which ensures that corporate and operational procedures are integrated to guarantee that the Halifax office continues to deliver its core services within ABC Transport.

These examples are simplifications intended to show that the mission statement for each of these levels must be appropriate to its level in the organization and are representative of the mutually understood and supported overall expectations that the organization places on them.

3.22 How Does the Mission Statement Fit into Critical Infrastructure Protection?

Though this chapter has looked at security as a primary role, the role of security within the organization, as defined within this chapter, dovetails well into the critical infrastructure protection role. The traditional view of security is one aspect of critical infrastructure protection, as this becomes gradually clearer throughout this book.

Though national authorities may identify certain personnel, assets, facilities, information, or assets as being critical to the safety, security, or economic well-being of its citizens (critical infrastructure), this is only one aspect. Within the context of trade and commerce, those that are providing transportation services or are supporting them must provide inputs into the performance of a system that must be available in case of catastrophe to avoid a significant impact across society. In the grand scheme of national exports, the loss of a single port or rail line may or may not be significant. The loss of the ability to maintain a level of throughput to maintain trade may be catastrophic given just-in-time systems and the need to maintain the flow of goods and resources using the system.

In an organization, the mission statement provides the clear intent of the organization with respect to what service or good it intends to provide. At the tactical level, this becomes the raison d'être for the organization. At the operational or strategic level, the assurance of personnel, assets, facilities, activities, and information is necessary to avoid disruption throughout the overall system.

3.23 Questions

Please note that the answers provided give one possible answer for the method queried.

Given the organization that you work for, or an organization you participate in, how would the concept of strategic, operational, and tactical levels fit into its structure?

- This describes the early part of the chapter. Head offices act as the strategic level within organizations whereas regional (district) offices act at the operational level. Storefronts and front-line service delivery entities act at the tactical level. It should be clear that the corporate headquarters will have elements of the strategic (location of senior decision makers) and tactical (installation protection) encompassed within the same entity.

Though strategic, operational, and tactical levels are relatively interdependent, they are set to function within frameworks and structures defined by the level one higher. Why is this necessary in a just-in-time system?

- The desired outcome provides a structure in which the right entities and appropriately delegated people make the right decisions with the right information at the right time in the process. Involving all levels of an organization in every single decision would likely cripple the organization as it would be overwhelmed by the number of decisions that have to be made.

Though the operational level describes the sum of its parts, why is it more appropriate to describe the operational level as the capacity offered by the sum of the entities at the tactical level?

- Individual capacities offered by an installation do not actually reflect capacities within the operational level's context. It is the arrangement of these capacities into useful structures at the operational level that define the operational level. For example, two facilities could meet the full requirements at the operational level, but if they cannot exchange demand (e.g., not connected by routes), then the capacity of the system is less than if they were. This is discussed more in the section on redundancy and resiliency.

Notes

1. The convention establishing the IMO was adopted in Geneva, Switzerland, in 1948, and the organization first met in 1959. The main task of the IMO has been to develop and maintain a comprehensive regulatory framework for shipping, and its remit today includes safety, environmental concerns, legal matters, technical cooperation, maritime security, and the efficiency of shipping. A specialized agency of the United Nations with 167 member states and three associate members, the IMO is based in the United Kingdom with around 300 international staff; http://www.imo.org.
2. The consequence of the studies initiated by the United States and subsequent consultations among the major allies was that the U.S. government extended an invitation to 55 states or authorities to attend in November 1944 an ICAO conference in Chicago; 54 states attended the conference. At the end, a Convention on International Civil Aviation was signed by 52 states to set up the permanent ICAO as a means to secure international cooperation and the highest possible degree of uniformity in regulations and standards, procedures, and organization regarding civil aviation matters. The International Services Transit Agreement and the International Air Transport Agreement were signed at the same time; http://www.icao.int.
3. The ownership and management chain surrounding any ship can embrace many countries, and ships spend their economic life moving between different jurisdictions, often far from the country of registry. There is, therefore, a need for international standards to regulate shipping—which can be adopted and accepted by all. The first maritime treaties date back to the 19th century. Later, the *Titanic* disaster of 1912 spawned the first international Safety of Life at Sea (SOLAS) convention, still the most important treaty addressing maritime safety; http://www.imo.org.

4. http://www.dot.gov.
5. http://www.tc.gc.ca.
6. For example, reviewing the Transport of Dangerous Goods Regulations (Canada) and 49 CFR (Code of Federal Regulations, Title 49, Transportation) for the United States, we find a number of restrictions relating to the movement of dangerous goods within each respective country. These regulations address the movement of dangerous good through populated areas and through areas where a significant risk to public safety occurs. The Controlled Goods Directorate of Public Works and Government Services Canada (PWGSC) places a number of limitations on the transfer of certain items (http://www.cgp.gc.ca/cgdweb/text/index_e.htm). This organization, however, is one of many organizations in both Canada and the United States in which the primary focus is ensuring that items or goods that may pose a risk to citizens do not fall into inappropriate or unauthorized hands.
7. The term *appropriate use* means using methods to impose restraint or resistance at nominal levels as preemptive security countermeasures, without imposing too much delay or too many inefficiencies that would otherwise degrade performance.
8. This is often referred to as a *service-level agreement,* which is a legally binding contract that stipulates certain levels of service and workmanship conveyed between operational and tactical levels of management.

General Definitions and Approaches

4

Objective: This chapter will:

- Explain general definitions used throughout the book.
  - Define several terms used in a way not normally used for the transportation sector.
- Provide definitions of terminology used within conventional security regimes and other risk-driven regimes that have been transmogrified to include frameworks within which critical infrastructure protection operates at this early stage.

4.1 Introduction

Critical infrastructure protection (CIP) challenges traditional concepts of organizing programs into administrative silos. Many professionals practicing in the field will recognize the challenges faced as programs intersect and, in some cases, even come into conflict. The fact is that security incidents are only one potential source of disruption to critical services. Critical infrastructure protection demands that professionals set aside any issues they might have and focus on finding solutions to a number of challenges. Consider the recent situation in Denver, Colorado, in December 2006 that left thousands of passengers stranded, or another example involving passengers at Heathrow Airport in London (at the same time), which estimated to have affected more than 40,000 persons.[1] The concern focuses entirely on recovering travel capabilities (resuming a status quo of the system). The issue of how to deal with extended periods of an interruption or disruption will follow later.

4.2 Persons, Assets, Facilities, Information, and Activities

Looking at delivery of services, many rely on processes functioning as intended. These processes are supported by elements of personnel, assets, facilities, information, and activities. Whenever elements within the system are not capable of performing as anticipated, their levels of disruption can easily lead to a loss (people, property, or both). This topic discusses further relationships in Chapter 6, but it is important to note that the performance of the system relies on specific performance factors (by processes and their component elements) and how those components are coordinated to their best use.

For example, look at the composition of our fictitious company ABC Transport Company. Our example will find a number of personnel (e.g., management, administrators, planners, drivers, maintainers), assets (e.g., computer systems and networks, vehicles), facilities (e.g., warehouses and coordination centers), information (e.g., pick-up points, destinations, billing information, routing), and activities (e.g., loading, unloading, trans-shipping, monitoring and tracking) supporting the processes that drive the business effort (mission) forward. This list can get rather extensive, as primary factors rely on nonprimary factors, which create interdependencies. Understanding how these components and their supporting nonprimary factors contribute to a cohesive system (collection of processes) is of primary importance to CIP (including security) managers and practitioners.

4.3 Follow-the-Pipe Approach

This concept describes working back from a final product to identify the inputs previously described. The term *inputs* describes persons, assets, facilities, information, and activities that contribute to a process working toward achieving the final product. This is similar to processes used by those involved in process mapping, process flows, or conducting process analysis. For the nonprofessional, this may be answered by asking, What do I need to meet my goals? Many organizations begin with an end deliverable and then describe the primary level of inputs required to meet that deliverable. Having established the primary processes (and the inputs), organizations then examine the processes necessary to maintain those inputs. By following this approach, anyone following given inputs to the pipeline can identify single points of failure (something to avoid in any process) as well as key elements that must be protected from disruption should the level of performance need maintaining.

Understanding when to stop deconstructing a process is an exercise in judgment that requires some analytical skill and thought. A general rule of thumb is to go back until there is such adequate redundancy in the system that disruption is no longer a significant factor. Consider the need to maintain the ability to find appropriately trained and capable information technology (IT) college network administrators. Most network administrators usually work backward in a manner that provision the college program that educates network administrators factored into the equation. This, of course, goes beyond the threshold of being reasonable, as the organization can just go to another college to find a graduate who can still be put through "on-the-job" training to make him or her familiar with the network's specific quirks. It is a reasonable case to argue that the organization's decision only to have one network administrator (due to cost) could pose a risk to the system, particularly if that network administrator becomes unavailable for any of a number of reasons (Table 4.1).

4.4 Mission-Driven Value

As discussed earlier, there are two major priorities at work within the system. One priority, which regulators generally focus on, involves ensuring that the transportation system is not used inappropriately to facilitate an attack or similar kind of activity. How regulators influence the private sector in addressing this risk is discussed later in this work. The other priority, from the point of view of private-sector management, is the protection against loss such that business activities remain viable and profits generated (or losses minimized). In this case, we are talking about assurances of infrastructure in terms of the persons, assets, facilities, activities, and information and the value that is based on how they contribute to the final product or the ability of the system to perform.

This speaks about value of the transportation system in how it supports society by providing services that many other sectors are dependent. Assigning any value based on this criterion indicates the real purpose of CIP, which involves the protection of an essential service or good in such a way that the public at large can expect the sector to be able to maintain tolerable levels of performance on demand.

For example, ABC Transport's mission provides timely deliveries of parcels requiring that certain parts of the infrastructure are present and functioning while other parts play a less important role. One critical requirement involves the ability to plan the movement of the parcels, even during periods of disruption. Senior management therefore will likely allocate more resources to meeting this requirement than to office-cleaning activities to maintain its comfort zone.

4.5 Vulnerability-Driven Considerations

In the case of the transportation system, we must also consider that the very service that it provides is used as a mechanism of delivering an attempted attack.

Table 4.1 Examples of Inputs

Service	Tier 1	Tier 2
ABC Transport Planning of a Shipment		Persons: not applicable
		Assets: computer for booking
		Facilities: planning center
	Persons: route planner	Information: planning procedures and daily situation reports for routes
		Activities: entry of information into the system
		Persons: maintenance personnel
		Assets: connection points to system
	Assets: booking system server	Facilities: server farm in building
		Information: maintenance and troubleshooting details
		Activities: maintenance of environment
		Persons: facility management, security
	Facilities: server farm with booking system server	Assets: systems maintaining environment
		Facilities: not applicable
		Information: roles and responsibilities
		Activities: maintenance of environment
	Information: pick-up point, destination, required arrival time, route (generated by system and planner)	Persons: database administrators
		Assets: computer interface
		Facilities: server farm
		Information: as listed
		Activities: ensuring details are correctly entered
		Persons: planner
		Assets: routing board or status board
	Activities: planning of the route and integrating the route into work instructions	Facilities: operations center
		Information: database holding key information
		Activities: coordination with supervisors who are directing daily tasks

As a result, it is important to understand the value of the system in the context of how various components within the system are exploited to fulfill or further an attack or how they create conditions that are conducive to an attack.

Regulations address these issues by creating constraints on the system. These constraints followed for the organization considered as conducting "legitimate" operations. Generally, they are based on an evaluation of risk and perceived vulnerability in the system determined by regulators. The specific risk mechanism used in this context is discussed later in the chapter.

For example, ABC Transport drivers that are going to be handling dangerous goods must also undergo background checks before they can be assigned driver responsibilities where dangerous goods are involved. Although ABC Transport conducts its own background checks on its drivers as part of its human resources program, the real reason for a background check stems from regulatory requirements. Periodic inspections of records ensure that background checks are performed. If ABC Transport fails to meet these requirements, ABC Transport faces fines upward of $25,000, periods of up to one year in prison for management personnel, and a suspension of its operating license for a period of up to six months, depending on the severity of the offense.[2]

4.6 Integrating the C-I-A Triad

The concepts of confidentiality, integrity, and availability (C-I-A) acting together form a solid, fundamental, operational practice model. For those within the security domain, these concepts are universally accepted and are relatively common terms used to identify the assets process. These characteristics are applied against the personnel, assets, facilities, information, and objects to determine specific contributions they make and what kind of damages would result to the organization should those characteristics somehow be compromised.

4.6.1 Confidentiality

This bases the need for control of information within a trusted community. As an example, business plans identifying new market opportunities are valuable in terms of potential opportunities of which competitors are not yet aware. Essentially, advantages provided by confidentiality involve one organization having clear information while other organizations are still subject to the "fog"[3] associated with not having that information.

Another aspect of confidentiality is that certain information may be of significant value to a hostile party and may, should it be revealed, become a detriment to its originator. For example, a security plan, though valuable to a business, would give a potential attacker information with which to plan or coordinate an attack if it fell into his or her hands. The plan itself has lost its value, in terms of coordination, and assumes a "negative value."[4] By describing the measures in place, the plan can then serve as a map to defeat the system. This negates much of the value of the security measures that had previously been implemented.

Within the transportation system, confidentiality factors into the equation rarely. Generally, confidentiality does not pertain to crew lists or other lists unless such information used to identify persons of interest, such as a head of state. Where assets and activities are involved, confidentiality generally factors into efforts to certify that it is difficult to identify specific capabilities of systems (e.g., the kinds of security equipment used to guard a facility or security planning details).

4.6.2 Integrity

Integrity focuses on whether or not something is performing as it is intended or designed or can be trusted without requiring further verification. This generally uses the context of information management (IM) and communication systems where changes in the data or messages received can have an impact on the performance of specific tasks. For example, if a system requires an e-mail to communicate with its clients, the records holding that address information become very important in making sure that the client receives the right information on time. Similarly, navigation systems that cannot guarantee that the latitude and longitude that they indicate (to the user) are, in fact, accurate to a reasonable extent are quickly discarded as being of no use.[5]

The integrity of personnel, although perceived as a difficult concept, can be reduced to two major factors leading to a level of confidence regarding whether or not an individual will perform as anticipated. First, does the individual pose a threat to the system in terms of wanting to cause it harm (terrorism) or to use it inappropriately (organized crime)? Security screenings for employees and similar activities generally address this. Second, can the individual perform the role or provide the function expected by the organization? This has traditionally been the role of human resources and the hiring agents to identify but can factor significantly within the CIP domain.[6]

The integrity of assets involves how close an asset or object aligns with its design specifications and whether or not can be trusted to function as intended. Part of this is obtained through quality assurance processes within the procurement process. Other aspects might be found in safety management systems, which examine conditions under which something is used and whether or not the asset is being maintained in such a way that the risk of its failure is clearly understood and predictable.

In the transportation system, there are several systems in which the integrity of information or processes factors extremely highly. Navigation and beacon systems, particularly in systems providing guidance in higher risk areas or conditions, must provide accurate and timely information to mitigate risks associated with navigation (e.g., differential global positioning systems, beacons, instrument landing systems).

Integrity aligns closely with quality assurance when looking at the performance of work. Consider, for example, the maintenance of unique or particularly valuable equipment. This kind of equipment is generally targeted for preventive maintenance due to its value. This maintenance may consist of a number of checks and inspections to ensure that there are no indications that a fault is developing in any of the systems used by that equipment. If the maintainer of the equipment begins to cut corners or develops cheats for the maintenance of the equipment but continues signing off on reports that equipment is at tolerable safety and operations guidelines, this opens up the organization to a layer of risk associated with unexpected failures or losses, especially if the equipment causes catastrophic damages.

4.6.3 Availability

This speaks of something or someone being present when demanded as part of a process moving forward. Consider, for example, the electrical power supply. If operations require electricity to function, availability of electricity is essential to a functioning system. It does not necessarily matter where the power is coming from (it could be a generator or part of a power grid)—only that there is electricity to power the system. When looking at operations within the transportation sector, there are a number of these kinds of services provided by personnel, assets, information, and activities, and they function at a variety of levels (i.e., tactical, operational, and strategic). When the follow-the-pipe[7] approach is used, the CIP or security manager should be careful about examining the process not only for these but also for any qualitative checks that may allow the process to continue.

The availability of personnel lends itself to three kinds of personnel roles:

1. Those who hold special delegations must indicate that all conditions are met to their satisfaction and must have the authority to give some kind of approbation such that next steps proceed in a process. Within the transportation system, these may include specially delegated managers, operations managers, safety officers, inspectors, harbor masters, and pilots.
2. Those that hold special knowledge or skills must be present for certain operations to be undertaken. These may include the operators of special or unique equipment, pilots, captains, engineers, or the operators of specialized equipment (e.g., communications, RADAR).
3. Those belonging to specific groups or organizations must be used to per- form specified tasks or undertake certain roles (e.g., labor, trade associa- tions, guilds). These groups can often undertake decisions as though they were an entity, particularly when dealing with, for example, the presence of certain conditions that may not be of benefit or may pose a risk to their memberships.

The availability of assets focuses on the ability to use those physical objects as intended. Asset availability has three equally important aspects: (1) the physical presence of assets; (2) the ability to function; and (3) the ability to function legitimately. Consider a ship being required to move containers. Three questions should be asked:

1. Is there a ship tied up that can receive containers?
2. Is the ship able to receive containers and hold them safely?
3. Are there sufficient items on a ship to prevent movement of the ship (e.g., lifeboats) if they are not present or in operable condition? It is not that the ship cannot function without a lifeboat but that the ship cannot legitimately function without that equipment and the crew, noting its absence, would likely disapprove of the condition.[8]

Some view physical facilities as assets per se; however, looking at availability of facilities, the all-hazards approach can apply. Should the organization's key assets be held within a facility and require persons to use them, the inability to access the facility causes a level of disruption similar to the loss of the assets. This disruption may take a number of forms, ranging from physical attacks against the structural integrity of the installation (e.g., car bombs, truck ramming, fire) to labor disruptions that prevent individuals from being able to enter or even the loss of key infrastructure within the facility that would require that the facility be closed for occupational health and safety issues.

The availability of information is of particular interest, especially when paired with the availability of certain kinds of systems. Consider the puzzle of finding a needle in a haystack, and translate this to finding a specifically numbered container in a field of thousands, if not tens of thousands, of containers. Suddenly, the availability of the yard plan and the information within it becomes very important, especially when looking at preventing any delays. Similarly, the availability of navigation systems, particularly in narrow or congested waterways and flight paths, is also just as important. Many of these systems are outside the control of the organization itself (except its own assets purchased to receive the signals). This also raises the question, in this electronic age, of how many persons are adequately trained on mathematics and similar skills necessary to navigate without the aid of those electronic systems.[9]

4.7 Integrating the D-M-L Triad

Looking at this process means inverting the C-I-A triad model into perceiving how values are translated to a negative impact. Where C-I-A refers to the confidentiality, integrity and availability of personnel, assets, and information, the D-M-L model refers to the disclosure, modification, and loss of those same resources.[10]

This pertains, as relevant, to each of the categories—personnel, objects, facilities, information, and activities. Of course, certain aspects of the D-M-L triad are more relevant than others, and it should be noted that more than one aspect of the D-M-L triad might apply to each category. For example, the processing of information will rely heavily on the ability to maintain integrity and availability of information. In certain aspects, confidentiality of information also factors into this consideration. What needs to occur is an understanding of how the D-M-L triad operates with relation to those five categories.

4.7.1 Disclosure

This may be viewed from several angles. First, what characteristic makes information sensitive? Second, look at the nature of the unauthorized characteristic. Although a manager may declare that certain information may not be disclosed, the manager is actually working within the delegations and requirements set forward by the organization's higher management. This is a difficult topic dealing with the concept of values and ethics when looking at assurances of critical infrastructure, and such questions are worth considering given the nature of the possibilities of multiple impacts.

Loss in terms of confidentiality is based on some kind of advantage and is realized as being the only holder of that information. Once information has been released or suspected of release, there is a question as to whether or not the information continues to retain any of its primary value.

4.7.2 Modification

In contrast to disclosure, modification deals with whether or not the information can actually be trusted. Consider a navigation system scenario again. At the client–user interface, the core need is for information to be accurate and current. Looking at how information is generated finds a number of questions requiring answers, which might include the following:

- Is the concept sound?
- Are data generated accurately?
- Are data communicated the same as data generated?
- Can data be degraded (either themselves or through corruption of their carrier signals), intercepted, or mishandled?
- Is the receiver subject to any sort of data-corruption issues, such as in certain kinds of operating environments?
- Can the end user read the data?

Modification also deals with conditions where something is changed in such a way that there is a greater probability of a different outcome than anticipated. Within the D-M-L context, this is viewed in terms of a negative impact. For example, fuel sources can be contaminated. Machines might be mistuned or modified so that they break down where, in fact, they should have performed. Similarly, processes might change by cutting corners during rigorous inspection processes, leading to a reduction in confidence associated with the system conducting the inspection as well as to an increase in opportunities for potential points of failure remaining undetected.

4.7.3 Loss

This describes a condition where something that needed to be available is lacking, and, as a result, operations are either halted or hindered. In some cases, loss may be caused from an attack. Sometimes internal routines of an organization may lead to condition of loss of some systems. Consider maintenance on the file-sharing server that leads to loss of service. If the organization relies on that service, then maintenance, although necessary, has contributed to a condition of loss. Most organizations have a sense of this and schedule maintenance activities during off-peak hours when losses associated with disruptions are minimized. However, for organizations that have 24/7 operations, there will always be some form of impact.

Disruption is considered a form of loss not permanent in nature. In this book, we look at loss as being independent of permanence and focus rather on the effect that the lack of a particular person, asset, facility, or piece of information or activity has on the performance being considered.

Continuing the example of ABC Transport, the planning system of the company feeds various systems that direct drivers where to go and tells facilities what to expect. As a result, the loss of this system's availability would have a significant disruption in the area where service was lost, and there would be no way to determine what steps were to happen next in the delivery cycle. Additionally, data within this system are trusted by the users or have integrity without performing checks, meaning that the corruption of the data (i.e., modification) could send personnel to the wrong location or cause the shipments to be scheduled past the delivery date.

4.8 CIP Management Approach

In such instances, most managements require an "all-hazards"[11] approach. Many of the risk-based entities focus on certain kinds of events or causes as they approach the issue from preventing something from happening. For example, safety management systems try to guarantee that incidents do not happen by following a set series of processes and procedures, working toward engineering risk out of the system. As the process moves along, the chances of an event occurring are gradually reduced until at such a time the risk is acceptable to both the organization and regulator. In CIP, attempting to reduce the risk of any disruption (loss) of the service provided means that any source of disruption may come into conflict with the CIP security manager's mission.

For our continued scenario, ABC Transport has a wide range of issues to consider when looking at potential disruption of the planning system or the modification of its data. On one hand, issues such as electrical power outages or brownouts due to weather conditions are beyond the control of the facility. Employees must undergo training and testing on the system (to prevent accidental modification of data) before they are allowed to access it and can only have unsupervised access to the system after they have undergone an additional review of their previous year's performance and a criminal record check. Similarly, procedures within ABC Transport only allow access to the system for those that need it, and as soon as that need is no longer present, access to the system is canceled, such as when an employee is terminated.

4.9 Criticality

This generally refers to how important an individual, asset, facility, piece of information, or activity is to a process, being careful not to confuse the concept of criticality with impact. Though this will be discussed later in more detail, the key element that divides the two generally deals with the element of time and how time and criticality, when considered together, impact the performance of an organization.

For example, the planning system, as described already, may be interpreted as a critical system as it has an immediate and significant impact on the organization's ability to perform. Other information management systems may not have the same level of impact (e.g., archiving server), making them less vital to the progress of work. Approaching this might look at the effect of the injury in terms of "zero time"[12] and if work could actually progress without component failure.

4.10 Means, Opportunity, and Intent

Looking at the probability of an attack is really looking at interaction among intent, means, and opportunity afforded to an attacker. Identifying these requires facility management to be in touch with various law enforcement agencies (or their points of contact) and may involve some candid dialogues regarding operations and intents of the organization.

The first aspect of this involves the concept of intent. This may be simplified in terms of determining whether the organization is of potential interest to an attacker related to its symbolism or potential disruption.[13] Understand that the attacker may look at a tactical-level target from a strategic level, choosing to attack a nation or its confidence by going after a local target where the shock value may be of particular value to the attacker's desire to cause fear, to instill panic, or to cause disruption. The attacks of the Madrid train station, the tubes in London, and the subway system in Tokyo provide examples.[14] At the other end of the spectrum, the management responsible for protecting CIP must take into account hazards that do not necessarily have any intent, such as natural disasters.

The second aspect involves the concept of means. When conducting an attack against an installation, the attacker may require special skills, knowledge, abilities, or tools to exploit vulnerabilities in the system. Does the aggressor have to bring these with them, or can they be found in the environment? In this context, the management will have to exercise a level of creativity, ruthlessness, and cunning. In some cases, the means of carrying out an act of violence may be integral to operations in this area, such as the presence of welding materials at a facility. Within the context of CIP, the question is not whether or not the means of an attack may be removed (which, incidentally, is not feasible) but whether various measures raise complexities of an attack. Costs associated with safeguards must be in an appropriate balance with the likelihood of a potential attack, given the knowledge, skills, and abilities an attacker would need to possess just to be successful. This is accomplished through the use and configuration of safeguards or through the reduction or realignment of vulnerabilities. Remember that risk may only be reduced—never removed.

Finally, the third aspect is the opportunity given to an attacker. As with any challenge associated with the means of foiling an attack, management must be relatively creative and must exercise a level of sane judgment. It is not feasible, particularly within the transportation system, to expect that all opportunities to carry out an attack can be eliminated. Consider the movement of small vessels in a port—simple to control until recreational boaters and other similar activities are factored in. Also, consider the sheer size of the rail network: Can every foot or meter of track and tunnel be watched? What is needed is an understanding of vulnerabilities in the system and installation of physical, technical, procedural, and psychological safeguards such that management can respond flexibly to those vulnerabilities.

4.11 Convergence within the Transportation System

For the CIP security manager, the transportation system poses a number of challenges. Given that the transportation industry operates with a high degree of convergence among physical, information management, and technology systems (e.g., scheduling systems, planning systems, navigation systems), the CIP security manager must have a general understanding of the ambient threat environment associated with physical and information technology as well as management environments. Given that transportation systems (similar to our previous analogy of human infection) can be attacked locally to accomplish strategic intents (e.g., an attack against mass transit systems or key nodes), the CIP security manager must be able to generate capabilities (e.g., permissions) that exist within the system and how they are distributed.

The key to addressing this issue lies in the ability to draw on the expertise at all levels of the organization as well as to be in contact with communities of experts. Going back to the analogy of infection, the CIP security manager will need to understand the conditions at the tactical level where failures are most likely to become apparent and how the impact of those failures will ripple throughout the system. This will provide the CIP professional with an understanding of how the operational and strategic levels are eventually infected and to what extent.

Communities of experts can provide valuable data and insight into the potential vulnerabilities of systems and, if those vulnerabilities were exploited, the nature of the impact at the tactical, operational, and strategic levels. This applies not simply to organizations that deal with the IM/IT[15] issue but also to all operations within the organization.

In looking at potential losses because of disruptions in of ABC Transport's planning system, it is also discovered that the planning system records whether routes and tasks have been completed. This information is fed into the employee pay system nationally, due to the high use of contracted drivers that remain resident in the system, and will not process their paychecks until it has determined whether a driver has been active during that period. As a result, a disruption in the planning system could, if not properly followed up, actually create a situation in the pay system resulting in unpaid employees.

Similarly, the planning system also keeps track of vehicle mileage, feeding that information into the maintenance scheduling system. This books specialized mechanics and arranges for space at local garages with which ABC Transport has arrangements. A disruption in the planning system has to take into account any unrecorded mileage that has to be entered manually into the maintenance system so that vehicles are maintained in accordance with various municipal and other regulations and bylaws.

4.12 The Concept of Risk, Residual Risk, and Risk Appetite

Risk, generally defined, is a condition of uncertainty regarding the possibility of loss.[16] The concept of possibility speaks to the means, opportunity, and intent discussed previously. The concept of loss is a bit more challenging to describe as it focuses on three aspects:

- The first aspect is specific to the organization within which the entity operates, such as the private sector. Categorization of the loss is generally based on the tolerance of the organization to accept such loss.
- The second aspect deals with a level of liability that the entity could assume because of losses that occur outside of the organization, such as damage within a community. In most circumstances, the tolerance may be described as a combination of the losses associated with the liability (e.g., dollar value and credibility) but may also be driven by society's tolerance for risk, which is communicated by legal mechanisms, such as laws and regulations.
- Finally, there is the loss associated with the performance of the system in general, and it is defined by how much capacity over what time period will no longer be available within the transportation system.

In this context, the first, organizational loss, sits firmly within the domain of security and loss prevention. When dealing with the impacts on society, some begin to move into the realm of emergency preparedness (encompassing emergency response, emergency preparedness, and disaster relief). Critical infrastructure protection encompasses these two while also placing an emphasis on the third.

One has to be cognizant of missions associated with various levels of the organization as those missions form part of the foundation that builds risk decisions. Most facility security officers are familiar with the tactical level. Here, the primary importance is the ability of the facility to survive and maintain its position within the organization. At the operational level, the value of the facility is closely tied with the capacity that it affords the entire system. If the facility provides a key link between numbers of conduits that play a significant role in terms of capacity, the priority may be placed on it above other facilities (or the opposite). The strategic level of the organization may also consider this but will also interpret the events and risks present in the facility against the strategic mission of the organization. Finally, regulators (e.g., Transport Canada, U.S. Department of Transportation) need to understand that facilities may take on some aspect of national importance, depending on goods or services offered and how they align with national priorities (e.g., trade, support to a region).

The term *residual risk*[17] follows similar patterns of risk in that it also changes as one moves among tactical, operational, and strategic levels. Residual risk refers to perceptions of how much risk remains within a system after the organization responds to the discovery of risk in the system. In some contexts, this is communicated in terms of residual risk being the amount of risk remaining in the system following the consideration of safeguards and vulnerabilities in the system. The first statement is used more within this work, since the latter implies that a level of action is taken on discovery, something that does not necessarily reflect the challenges associated with ignoring risk.

For example, ABC Transport may discover that its facility in City A is at extreme risk of a shutdown due to disruptions at a neighboring facility that could close the road into the warehouse. While that particular facility may decide that the best course of action is to either establish another route or take some kind of action to make certain that the route remains open, the operational level may decide that it is better to establish a second warehouse away from that area or even to relocate the warehouse altogether. These are both strategies as to how to address the discovered risk, but both organizations have different perceptions about how much risk should remain in the system (i.e., residual risk) following their decisions.

This perception focuses on the concept of *risk appetite*,[18] which also functions differently at all three management levels. Once again, this refers to mission statements associated at each layer of the organization. While the tactical level may choose only to accept risk that preserves the specific site, as demanded by its role in the organization, the operational level may decide that there is no way for any action to be taken at the local and tactical levels. As a result, a conflict between the local (tactical) and regional (operational) management could arise as each attempts to meet its own obligations to the organization but to approach the challenge from within their divergent mandates. Finally, the strategic level may argue that as long as it does not cost an inappropriate amount in terms of resources or lost potential, the situation is tolerable. The strategic level may simply accept whatever decision is made as long as those two conditions are maintained.

ABC Transport's planning system has been declared a high-value system and currently operates in isolation with no backup system or redundancy. Senior management, concerned over the impact that the loss of the server could cause, has decided that it must be assured that the server's functionality cannot be lost except under the direst of circumstances. The manager responsible for IM/IT has established a system in two of the regional control offices that include a mirrored server that would take over in order of precedence should the primary server fail. As an example, an electrical power failure in Boston would mean that the server in Halifax, Nova Scotia, would take over the functions and, should that server fail, the system in Seattle would then take over, making the system impervious to all but most major systemic events. Senior management has examined the costs associated with the proposal and balanced them against the potential impact associated with the single server and has decided to allocate the operating funds necessary to maintain the system with redundancy, since risk is now tolerable.

4.13 Who Decides the Threshold for Risk Appetite?

Ultimately, this question will lead to three different answers depending on who is accepting the residual risk and on behalf of what organization or entity.

Recapping risk at the strategic level within an organization, strategic management must ultimately accept the risks associated with specific courses of action as they bear the accountability for the performance of that organization. This kind of approach migrates downward within the organization through operational and tactical levels where there are no significant impacts on any levels above them. In essence, the entity holding the responsibility and accountability for enterprise risk management (e.g., financial, labor, safety, environmental) becomes the entity that bears the accountability for the transportation system security risk management that they control.

In a similar sense, various governments associated with the transportation system have dictated certain boundaries on the risk appetite through the promulgation of legislation and regulations that constrain various operators. While an operator may choose to accept the risks associated with the security issues (for whatever reason, which could range from a lack of awareness to a perceived inapplicability of a specific issue), governments may decide that certain kinds of risks are intolerable across the entire system. Many people experience this system every time they fly and are informed that certain kinds of items cannot be brought aboard an aircraft. The use of regulations and directives allows the regulator to shift the risk away from the security issue and into the realm of operating (permission to legitimately operate or certification) and financial risk (fines and other forms of penalties) to maintain the attention of the target organization's management.

The final, and perhaps ultimate, risk appetite resides within the client base of the transportation sector. This is also referred to as public confidence.[19] Should it be perceived by the client base that the system is not secure, then the client base will simply withdraw, reducing its own risk exposure to the system as much as it can. Following the attacks on 9/11 in New York City, the subway bombings in London, and the train bombings in Madrid, this was a primary concern for both operators and regulators and raised questions about how to ensure that people would return to using the system.

4.14 Avoiding, Addressing, Transferring, Accepting, and Ignoring Risk

There are several different options available to organizations when they discover risk within their various systems.[20] The option chosen will depend on their impact within the overall system (enterprise risk management and cost-benefit analysis). Each of these options deals with how the organization perceives the impacts, consequences, means, opportunities, intents, and other factors associated with the risk.

4.14.1 Avoiding Risk

In most contexts, this involves eliminating risk by eliminating the source of the risk, or in more common terms, changing the operations of the organization. Within the transportation system, this may or may not be possible. In the private sector, there are limited (if any) abilities to compel private sector entities to remain in business. In fact, in a free market society, it is unlikely that such direction would be tolerated within the industry, if not society.

4.14.2 Addressing Risk

When an organization chooses to address risk, it is because it has recognized that there are levels of impacts or consequences that are unacceptable to the organization and that there is a sufficient combination of means, opportunity, and intent present in the system to make it worth allocating resources. This approach does raise a challenge that has been brewing within the security community for some time: the balancing of impact and probability as dominant factors. In efficiency-driven systems, the tendency is for probability to outweigh impact in all but a few rare cases. This approach tends to allocate minimal funds into the system. Following the popular consensus regarding 9/11, there has been a shift away from probability and toward the potential impacts and consequences of an attack. As those in management are well aware, the costs associated with this kind of approach can be quite high, if not staggering, to an organization. Although both these sides have some merit, some might reasonably argue that what needs for a balance between the two systems.

When addressing risk, many have made the decision to do something about it. Again, the different levels of the organization may have different approaches. Turning again to ABC Transport, the tactical level may commit resources to confirm that it could not fall prey to the same kind of theft again—such as installing alarms or hiring security guards. What is important is that the chosen course of action may not be the same at the operational level, which may choose to route particularly valuable packages to a nearby office that does not appear to have the same challenges. At the strategic level, the management may choose to wage its own publicity campaign in conjunction with its advertising agency, attempting to show how the company takes the issue and security very seriously or reinforcing a public message of success in opposition to a local message of loss. The key at the perceived addressing of risk is that addressing risk involves committing resources or effort against the risk and its issues to reestablish a comfort zone.

Addressing risk involves being able to support decisions within the context of *enterprise risk management*[21] or its more holistic risk picture of the organization. When addressing risk, the gross percentage of resources are found within the organization or set to be committed from within the organization, leading to a situation where other risks may be opened up. Within the context of CIP, this decision does not come lightly, as this will cause a conflict between securing the infrastructure versus assuring service levels within the organization. Although addressing a security risk may work toward preserving the infrastructure of the facility, opening the facility to a new avenue of attack or risk is not desirable when considering the mission statement involving the assurance of services.

4.14.3 Transferring Risk

This is very similar to the addressing of risk but involves an allocation of resources to another organization that performs the function. This may involve a contract for specialized services or the purchasing of insurance. Those responsible have decided that the impact, consequence, means, opportunity, and intent all exist in adequate force for some concrete response to be required (usually an allocation of effort) but that the mechanisms being used are slightly different. This could be because the nature of the risk to which the organization is exposed requires some kind of specialized knowledge, skills, tools, or capabilities that the host organization cannot effectively nor efficiently provide and sustain. Risks associated with a chemical parcel may involve a very grave occurrence in terms of impact or consequence but are sufficiently rare so that the financial risks associated with the high costs of maintaining a first responder within the organization exceed the financial risk appetite of the management being asked to provide resources. As a result, the organization then looks for more efficient systems, often within the private sector or other service providers, that can offer the same results from a risk management perspective but that pose less financial risk and fewer legal risks (due to the specialized skills needed and due diligence issues). Once that mechanism is found, some kind of binding instrument, say a contract, is put into force that shifts the issue over to that other organization. This refers to the concept of *transferring risk*[23] as the organization has made some other organization responsible for addressing the risk.

It is important to note that the organization seeking to transfer the risk is not removing itself from responsibility or accountability by doing this. For example, if a client sends a package through ABC Transport, which then subcontracts it to another organization that fails to deliver it, the client does not have an issue with the subcontractor but rather with the organization to which he or she has paid money for a service. Thus, the organization should be looking at the costs associated not only with the contract directly applied against the service but also with the monitoring of the performance of that contractor so that it does not open a new potential risk avenue.

4.14.4 Accepting Risk

In certain cases the organization may accept the risk. This occurs where the impact and consequence of the risk is sufficiently low and where the means, opportunity, and intent sufficiently unclear as to lead to an assumption that the exposure to the risk is well within the risk appetite of the organization. Consider the risks associated with the theft of the occasional pencil from the organization. Given the cost of the pencil, and the relative infrequency that it is going to be stolen, as compared with the costs associated with preventing every occurrence, management may well decide that the couple of dollars lost due to this is acceptable. Care has to be taken if adopting this approach. There should be very clear limitations established so that relatively minor issues do not escalate into something major—such as an individual deciding that if the company is willing to accept the loss of a pencil, perhaps the cost of a laptop will not be too much either.

Risk may only be accepted by the entity accountable for the risk or that would be held accountable for any impacts (including consequences) associated with the risk. For example, a neighbor may not accept the risk associated with the decision to install or not install smoke detectors in their neighbor's home. The impacts are borne by the family living in the home, not the neighbors.

The acceptance of risk is part of the daily operations of the business but is fraught with potential traps. Consider the risks posed by a new threat. Does management have a clear understanding of the risk to which the organization is exposed? In some cases, management may be making a number of assumptions based on experiences that are not necessarily indicative of the risk associated in that one particular case. Similarly, have the right levels of the organization been involved in the risk acceptance process? While the strategic level may decide that the risk is acceptable from an overall perspective, the tactical or operational levels may be constrained in such a manner that the risk cannot be accepted legitimately and must either be addressed or transferred. The acceptance of risk process must involve inputs from each of the levels to ensure that the final decision is valid and that the overall system remains aware of the risks to which the organization is exposed.

4.14.5 Ignoring Risk

Management may also choose to *ignore the risk*. This occurs when the CIP manager, or other managers, cannot convince management that the impact and consequence, means, opportunity, or intent warrant their attention and management makes the declaration that such an issue is not worth approaching. In an age where due diligence is becoming more and more part of the business decision-making culture, this is a very difficult approach to take in a risk-based environment as it can lead such management open to a range of potential actions.

For the CIP security manager, the ignorance of risk should signal an approach that involves determining why risk is being considered inconsequential. Does management require additional information, or does the management lack the willingness to approach a certain issue? Is it worth soliciting outside expertise (e.g., local law enforcement, legal support from within the organization) to present further supporting information, or is it worth allowing the issue to fade away? Given that the nature of CIP involves assurances of a particular service, it is very unlikely that the CIP security manager will find a management organization willing to accept the accountability and responsibility for openly ignoring a specific risk.

4.15 Responses to Risk and Regulation

Regulations communicate, in a variety of terms, the risk appetite associated with the overall system when looked at in the context of society in general. This is done by constraining the various organizations that fall under the regulations by declaring that certain actions or conditions must be met in specific circumstances, in essence limiting what can be considered an appropriate approach to dealing with apparent risk. This may be done by limiting the number of options available in terms of addressing, transferring, accepting, or ignoring risk or by stating that the risk appetite may not be appropriate below a certain threshold.

One of the challenges for the CIP security manager is to remain aware of the various constraints that are in operation around his or her area of responsibility. For the tactical level, this may be a reasonably short list. This list might encompass local, state, and perhaps some national constraints. As one progresses from the tactical to the operational and finally to the strategic level, this list of constraints becomes increasingly more complex and can become a factor in the planning of movement between areas.

ABC Transport, for example, ships certain larger packages by vessel along the eastern coast of the United States and into Canada. In doing so, ABC Transport has to assure that the U.S. regulations are met for the period that it resides within U.S. territory but also that applicable Canadian regulations are met when it arrives in Canadian waters. If it is discovered that the shipment does not meet regulations, it could be denied entry into Canada or delayed while appropriate assurances are given to the right officials. ABC Transport, understanding that it could lose its license to move these goods into Canada should it fail to comply with the regulations, takes steps as though it had chosen to address the risks present in the system by complying with the regulatory requirements.

4.16 Risk Awareness

This concept regards the organization's understanding of what risk it faces and, once again, operates differently at various levels of the organization. As with understanding the level of threat within an environment, this aspect of dealing with risk requires careful communication across all levels of the organization and in both directions.

From the tactical to the strategic level, the focus of communication resides with the building of awareness. While the concept of the transportation system is arguably more than a simple facility or entity, it is important to realize that the sum of the system is the sum of its components and their interrelational positions with each other. As a result, it is equally as important to understand the various risks that are present at each point and within each conduit of the system if one really intends to grasp the system-level risk.

This means that the tactical level must be constantly informing the operational level of situations that cause a change in the performance of the facility. This, of course, operates at the enterprise risk management level of the facility. The operational level must examine the information collected from the tactical level to determine if there are common areas of activity that are affecting the level of performance. This provides the foundation for systems to alert the organization to potential emerging threats, lessons learned, and best business practices. Once this information has been collated at the operational level and is assessed against the factors that are operating within that region, it is then communicated to the strategic level, which can then collate it with the information received from the other operational levels. This information is then evaluated to determine whether or not the information has value to any of the other components of the system (e.g., similar challenges, factors or influences in the system) in such a way that other parts of the system can benefit from the experiences and outcomes that have been reported.

From the strategic to the tactical level, the emphasis is more about establishing and maintaining a base on which clarity of the overview of risk is founded. When looking at communicating information across an entire system, there has to be a common understanding as to the meaning of that communication to prevent confusion. This involves setting standardized definitions and structures used to communicate information into the system. Similarly, the strategic level may also consider the establishment of reporting periods and procedures meant to certify that the information within the system is as complete as possible. This may involve making certain that the appropriate corporate policies are in place so that incidents are reported and that once they are reported, they cannot be closed until all the information required has been captured within the system.

This leads to three challenges for the strategic level. The first involves ensuring that the tactical and operational levels continue to report accurately events that affect performance. This involves the establishment of a culture that values the honesty and transparency of this kind of effort. The second challenge is understanding that, in this given threat environment, it is unlikely that all events can be predicted. In fact, that is the reason for adopting systems that take an all-hazards approach and that then base their capabilities on flexible and dynamic systems. Within the context of incident reporting and recording, care has to be taken to guarantee that any IM or technological tool does not simply reflect the threat environment during a snapshot in time but also can evolve in pace with its environment. Finally, the third challenge involves providing that the information is used to its best value. This often involves making certain that best business practices and lessons learned are used appropriately throughout the system in such a way that the information is meaningful at the tactical and operational levels. This also requires the establishment of a "thinking" culture and not a "cut and paste" culture within the organization, as the response that worked well in one environment may not be suitable or as effective in another.

ABC Transport has a policy that a driver will call in to a central number whenever it appears that the planning system information has not become available. The same individuals who allocate the jobs to network trouble-shooters almost instantaneously collate this information, collected from the tactical level, at the operational and tactical levels. A procedure is in place that should a high-value system (e.g., the planning system) go down, all lesser prioritized tasks are delayed until the problem is resolved. Progress tracking is done through the call center, which can provide updates to the various persons calling in.

4.17 The Concept of Safeguards

As with any risk, the concept of safeguards will change depending on the mission being considered (driven by the tactical, operational, or strategic location of the organization) and the environment in which it operates. Safeguards loosely defined, may be characterized by physical, technical (logical), procedural, and psychological measures that are evident within the organization that work to reduce the means, opportunity, or intent associated with an attack. This generally involves warranting that the chances that the preventive measures will delay an attacker long enough that the assault is stopped by various measures that prevent, detect, respond to, and recover from events. As a result, many look at both the level of the entity being attacked and the goal of the attacker in terms of creating a tactical, operational, or strategic level impact. For example, a safeguard may be applied to prevent localized damage (e.g., a fence, bollards), operational impacts (e.g., communication systems back to coordination points) or even strategic-level impacts (e.g., memoranda of understanding that come into force should a facility fail and items need to be rerouted). The next subsections look at this concept in more detail.

4.17.1 Tactical-Level Safeguards

These are intended to deter or delay an attacker. These kinds of safeguards focus on the protection of personnel, assets, facilities, information, or activities at the local level and may include such measures as fences, bollards, locks, or cameras.

Tactical-level safeguards work within the localized system to protect the infrastructure that allows that particular site to meet its mission requirements or goals. They are specific to the environment in which they are placed and are generally subject to the financial and management limitations of the facility, unless they can be linked to an operational- or strategic-level doctrine. The goal of these safeguards is to preserve, at that location, the ability to maintain a level of performance through providing a controllable and predictable environment in which work takes place.

ABC Transport has established a protocol where the driver must swipe his or her ID card through the system before the information is sent back to the planning system. This card is required to send any information from the device or request any information from the server. The device is set to power down within two hours of being inactive; this is done to prevent access to the system by an individual who may attempt to cause a system disruption or to corrupt the data in the system.

4.17.2 Operational-Level Safeguards

These play a different role. While the operational level may dictate or recommend a number of safeguards at the tactical level, remember that the mission of the operational level is to maintain a level of capacity across the overall system. Infrastructure associated with this role, such as command and coordination centers, may require tactical-level protection, but the safeguards associated with meeting the mission requirement also focus on the ability to detect changes in performance while understanding the options available to deal with those changes.

Operational safeguards work within the localized system to coordinate the flow and movement aligned with the various nodes and conduits in the system, shifting the flow depending on the capacities available. Within the strategic system, the operational system provides an indication regarding both the capacity of the system at an operational level while also providing a trigger for the strategic level indicating that there may be a system-wide impact developing and allowing the strategic level more time to respond to that potential event.

ABC Transport has established a protocol where, should a driver report in sick, a backup driver is called or the packages may be loaded onto other trucks when they return at noon to the central distribution hub. Part of the planning system identifies a primary option that could be used (based on time and proximity) should the planned option not be available for any reason.

4.17.3 Strategic-Level Safeguards

These may take on a number of different forms. At one end of the spectrum, the senior management of the organization might work to either increase or restrict a certain line of business, depending on that shift to maintain the performance of the organization. This may be reflected in a decision to promote a business line that is not subject to as high a level of risk as another due to an increase in threat. Concurrently, the strategic level may also enter into alliances or agreements with other entities for their mutual protection, such as an arrangement between two normally competing airports to handle each other's traffic if others are disrupted for whatever reason. In general, the strategic level will focus on how to maintain the organization's mission as best as possible.

ABC Transport also has an arrangement with a rail carrier that can assist should air travel be disrupted for whatever reason. Part of this system involves notifying clients of the delays that are likely to occur as a result of this shift and initiating a client-relations process to certify the satisfaction of the client base as much as possible. This plan has tactical, operational, and strategic level components that allow for the routes to be shifted, upon notice, given a disruption at any of the local, regional, or even national levels.

Should it become apparent that certain systems are being disrupted more than others, the decision has already been taken to allow for a shift to a more stable system as part of the routine conduct of business and does not need management to meet in its entirety.

4.17.4 Regulator-Driven Safeguards

Up to this point, we have been looking at scenarios that involve an organization attempting to protect its own interest. This only reflects the requirement to preserve the availability of the system. Another aspect involves safeguards that are put in place to prevent the transportation system from being used as a conduit, or delivery system, for an attacker or someone attempting to use it for illegal or nefarious purposes. While we have discussed, to an extent, the role of regulations, it should also be clear that these oversight bodies also hold the authority to require the imposition of specific safeguards depending on the kind of threat information they receive.

In these cases, the safeguard can operate at any of the tactical, operational, or strategic levels, the application being driven by the nature of the threat information being received. The regulator might require that all nodes within the system watch for something specific; as this might be reflective at a tactical level, it is a strategically scoped rule. At the same time, the regulator may require that organizations maintain certain capabilities as a minimum performance standard and clearly communicate their ability to meet those minimum standards. During disaster relief efforts, regulators could require that the system carry so much material for the effort within its own systems into the affected areas. Finally, the regulator can require that the organization only operate within certain constraints at all times, ensuring that no particular organization is able to become a vulnerability or risk avenue for a potential attacker.

4.18 Prevention, Detection, Response, and Recovery

As discussed earlier, the purpose of a safeguard is to deter or delay an attacker so that the attacker cannot successfully accomplish the attack. In looking at this statement, there are four elements to safeguards that can be put in place. Three of these—prevention, detection, and response—deal with the requirement to deter or halt an attack. The final element, recovery, deals with the ability of an organization to reestablish its level of performance.

The concepts of prevention, detection, response, and recovery are expressed in other forms. One common, and growing, approach involves the mantra "deter, delay, detect, deny, and detain." These approaches work toward the same goal of identifying appropriate safeguarding plans.

4.18.1 Prevention

Prevention deals with the ability to either deter or delay an attacker to such an extent that the assault is not successful. In this context, the deterrent effect stems from the apparent physical, procedural, and psychological measures being in force and visible to the attacker in such a way that the chances of success are not apparently in the attacker's favor.

At the tactical level, preventive measures generally involve the delaying or deterring of an attacker. These may include physical (e.g., fences), procedural (e.g., having to sign in), or even psychological (e.g., well-lit areas, reputation) measures that make it apparent to the attacker that the chances of success are greatly reduced. Success at the tactical level includes operations within that immediate area.

This approach changes at the operational level. The tactical-level entity focuses on its ability to maintain a level of service while preventing loss or an attack. The operational level has to remain concerned with being able to identify the status of its components so that it can appropriately route the movement of goods and services in such a way that the overall system's performance in that area is maintained. The operational level also has to be able to identify conditions within its area of responsibility that could indicate the risk is shifting into new areas as a result of imbalances between parts of the system and that offer an attacker the opportunity to cause the same impact. Consider, for example, the difference between a cruise ship and an international ferry where both vessels carry a similar number of persons and operate in a similar geographical area. If the system were to tolerate an imbalance in preventive measures so that cruise ships had many preventive measures whereas the ferries did not, conditions could exist that cause a shift of risk onto the ferries. This aspect is not limited to one particular mode, as the operational level must remain constantly vigilant and aware of the various intents of threat agents, aligning those with the potential opportunities within the system and communicating those to the strategic level.

The strategic level, in this context, focuses on being able to identify gaps in either the application of preventive measures (leaving vulnerabilities within part of the system) or shifts in risk across different kinds of operations within the system. Preventive measures associated with efforts at the strategic level focus on the system's ability to expand its capacity and to interrelate with all parts of its own system. Thus, it can link into other parts of the system so that the overall system becomes as aware as possible and has the necessary tools in place so that steps may be taken to minimize potential shifts in risk and opportunities to exploit their vulnerabilities.

4.18.2 Detection and Response

Should preventive measures fail, or appear to be at risk of imminent failure, each level of the system must be able to detect and respond to that potential failure in such a way that the impact associated with the failure is both minimized and contained as much as possible. It should be noted that detection includes the ability to identify a preventive measure has failed or that something suspicious is under way, and that the notification of the appropriate organization includes complete and appropriate details. Response, of course, depends on information provided (the reason notification must include a consideration of the appropriateness of the information) and the environment into which the response team will be arriving. Ideally, the time it takes to detect and respond to a potential event should be less than the time it would take a determined attacker to overcome the preventive barriers put in place to deter or delay the attack.

At the tactical level, detection and response deal with the protection of infrastructure in an immediate sense. Unlike at the strategic level, where the attack may still be theoretical in nature, an attack or detection of suspicious behavior at the tactical level requires an immediate and concrete reaction. This reaction focuses on first halting the event itself, making certain that any injury is contained, and finally assisting in establishing the conditions necessary for the organization to begin the recovery process. Concurrently, the tactical level has an obligation to report the change in condition to the operational level.

The focus for response at the operational level is somewhat different. Unless the tactical level indicates a disruption occurring, the operational level may choose to simply monitor the events, collating information for lessons learned and, ideally, intelligence purposes. Should a disruption occur, the efforts of the detection and response capability include the ability to detect the nature and scope of the disruption and then to respond by shifting operations in such a way as to protect the overall system and to afford the affected tactical level the time needed to reestablish control. Concurrently, the operational level should also be monitoring its area of responsibility to determine whether the attack is operating in isolation or is part of something larger while also keeping an eye out for any signs that the injury could be cascading into the overall system. As with the tactical level, once the event is determined to have had an impact that could affect the operational level, the operational level should be informing the strategic level.

At the strategic level, detection and response includes the ability to communicate the indicators of the events across its organization to either identify whether or not the attack is part of a larger pattern or forewarn other parts of the system of the nature of the attack. The response to this kind of event generally involves an assessment of the received information, communication of information to other parts of the organization, and communication of any special measures to be taken as a result of the event. Shifting of activities allows pressure to be taken off the affected area as much as possible and in support of the operational level, and the continued collation of information allows for lessons learned to be spread across the organization. Within this context, the strategic level is not leading the response to the events but is rather providing the necessary authority to decisions that are required to provide the best environment possible for those responding to the event so that the response and recovery periods are as brief and effective as possible.

The secondary role of the strategic level is also to ensure that the organization learns from the event. This may involve making sure that investigations or some other kind of activity takes place that determine why the system failed and what worked well.

4.18.3 Recovery

The final stage in the process, recovery, actually operates somewhat independently of this cycle but is essential from the perspective of the CIP effort. This stage involves the organization's ability to return to its normal operating levels and the reestablishment of stability necessary to maintain those levels.

At the operational level, the entity's report regarding its estimated time to recover to full operations is of significant value to the operational level as it allows the operational level to begin to assess when the system can be shifted back to its normal conduits and activities. Similarly, if the entity anticipates a time frame associated with a longer return to normal operations, when it is communicated to the operational level it provides a timeline that can then be used to begin to reestablish a sense of stability into the system.

The communication of this recovery is also of importance to the strategic level from the perspective not only of reestablishing normal operations but also of being able to demonstrate the resiliency of the organization. At the operational level, part of the recovery will involve dealing with any issues of public confidence associated with the event. At the strategic level, the strategic management and communications organizations will need firm data that allow for management to demonstrate that the event is under control and that there is no basis for further concern.

4.19 Looking at Vulnerabilities

Vulnerabilities may be loosely defined as the characteristics of the system that can potentially be exploited by an attacker or that reduce any ability to prevent, detect, or respond to events effectively. Although the security field generally does not include the second aspect of this statement, it is important to include it within the CIP domain as many of the potential hazards do not intend to exploit vulnerabilities; many organizations simply act according to their physical nature. Hurricane Katrina, for example, did not strike the coast of the United States with any malicious intent; it simply did what a hurricane is born to do.

When looking at vulnerabilities within the CIP domain, organizations must be aware that vulnerabilities operate somewhat differently from safeguards. Safeguards, within the CIP domain, operate differently at various levels of the organization and do not necessarily flow across from the tactical to the strategic level. The presence of a fence at a facility is of little importance to strategic-level planners as long as lack of a fence does not contribute to reduction in capacity of that tactical entity. Vulnerabilities, on the other hand, flow from the tactical to the strategic level due to their ability to be exploited and contribute to the failure of components of the system. The lack of a fence (e.g., indicative of a lack of perimeter control) may have a tactical affect (e.g., lack of perimeter control leading to persons gaining access), operational impact (e.g., difficulty attracting certain clients due to the apparent weakness), and a strategic impact (e.g., overall confidence in how the organization treats its security or protection). Each of these would require the organization to be able to prevent further damage and then to detect and respond to issues at each of the levels.

Similarly, the presence of a vulnerability that might be exploited is really an avenue that an attacker can use to accomplish her or her goals. It should also be considered as a systemic weakness that can be overwhelmed by hazardous conditions (e.g., a natural disaster, working environments). Should the attack be successful, the tactical level suffers the immediate impact, but the operational and strategic levels also undergo a kind of impact associated with the loss of capability and the need to adjust operations to maintain the capacity of the system.

4.20 Interim versus Proposed Measures

Within the transportation system, certain events will require an immediate response, which may be for any number of reasons, including safety, security, or environmental. Understanding the impact and apparent probability of an event could have aligned itself to such an extent that management has decided that it must act immediately to address or transfer the risk. Another reason could be that regulators have informed management that the organization is failing to meet a particular requirement for operations to continue legitimately.

Immediate responses will generally require the allocation of existing resources. Generally this will involve the addition of a new safeguard, the adjustment of existing safeguards, or the adjustment of the processes in which the vulnerability has been discovered to minimize the vulnerability. A lack of perimeter control and an on-scene security presence may be considered intolerable under a regulatory regime. As a result, the facility operator may choose to hire a guard company to provide the service or, depending on its operations, may adjust its operations and lighting to remove the vulnerability associated with someone being able to enter the yard undetected. These solutions are generally temporary in nature—at least until it has been determined what the preferred course of action is either to address or to transfer the risk.

Proposed measures deal with final decisions to commit sustained resources against a particular risk. This may involve courses of action to address the vulnerability such as removing the interim measure and replacing it with the proposed solution or adjusting the interim measure so that its adjusted form becomes the final measure.

From the CIP perspective, it is important to understand that interim measures will have an impact on other aspects of the organization, either in terms of the retasking of resources or of the impact on overall operations.

These impacts should be understood when the interim measures are imposed, and, if possible, there should be some timeline established indicating when the proposed measure will be put into force.

4.21 Layered Defenses

CIP deals with the maintenance of acceptable levels of performance, essentially ensuring that the required performance of the system will be there on demand (a question of availability). To accomplish or to facilitate this effort, reasonably careful safeguards put into force are actually appropriate in their environment. For example, if the goal were to protect a sensitive internal room at a college by building a 200-foot-high wall around the entire campus, the costs and impacts associated with the safeguard would be entirely inappropriate.

As a result, it is a logical next step to look at the concept of a layered approach. Remember, the goal is to increase the complexity of any attack against the infrastructure by requiring an attacker to possess more and more in terms of special knowledge, skills, and abilities. Similarly, most organizations want to be able to detect the signals that indicate a failure within the system as early as possible. The layered approach allows for a system to be implemented that provides a range or variety of measures to be used while also creating the necessary environment that supports the ability to detect potential failures in the system as early as possible. The first layer may require the visiting person to establish that they have legitimate business at the location, whereas the next layer may focus on the ability of an individual to present some kind of identifying token (e.g., an access card).

The concept of the layered defense also operates at the operational and strategic layers of the organization and within the system. Given an incident, the priority may be to contain the event within the locality but to increase the ability of those nodes and conduits attached to that locality to detect the indicators of both the event and the impacts associated with it. Concurrently, the operational level may redirect traffic away from the affected area in an effort to both reduce the pressure on the affected area while also reducing the chance of the impact entering the transportation system at another point. With the operational level communicating its actions to the strategic level, the latter may take steps to adjust the background activity of the organization in such a way that the flow of material into the alternate corridors is more graceful than it would be if redirected from within the operational area. Thus, the strategic level may initiate parts of its own response plan, such as communications or fan-out procedures, so that the capabilities are on hand and able to operate should the situation escalate any further.

4.22 The Macro Level

Up to this point, the focus has been on one organization's tactical, operational, and strategic levels. In reality, the system is composed of several organizations, representing a mixture of both publicly and privately operated sectors. As a result, one additional level of coordination and control must be examined. This level is evolving today and involves the coordination of the overall system by the government lead agencies.

For the private-sector entities, this would lead to a situation where the "lead" agency contains the relevant modal expertise but also the expertise in the manipulation of the system; although the span of this control is unclear, it should be noted that most CI sectors are regulated to some extent. Depending on the nature of control needed, the might reasonably find that modal organizations, working under the guidance of a system organization, exercise their various delegated powers in such a manner that the system is influenced to the system's best advantage during periods of increased risk.

4.23 ABC Transport

Consider that ABC Transport routinely delivers high-value parcels on a person-to-person basis, an increase in the number of attempts to gain access to senior corporate managers across North America has raised some concerns from major clients that the company could be used as a conduit for an event or even an attack.

ABC Transport's management recognized that the clients' concerns were worth looking into, due to the potential loss of those clients to more perceptibly secure organizations. As a result, senior management decided to demonstrate to its clients that security was a primary concern of the organization.

A review of the driver trucks indicated that many did not have doors that could prevent any but casual attempts at entry. Maintenance cycles were adjusted to include a better locking system for the doors while the parcel compartment was segregated from the driver and controlled by a lock-and-key system. Concurrently, ABC Transport's senior management issued a memorandum across its client base informing the clients of the concern and reassuring them that no ABC Transport driver would attempt to gain access through security but would recognize the mailroom manager's signature as the concluding signature if the company requested. Each client was asked to respond in 20 days through their local office with local offices updating the planning system with the new instructions.

At the same time, the company issued each driver an updated notepad that could electronically capture and display information regarding his or her routes. These notepads also included GPS locators (normal in the technology purchased) and software that allowed the driver to tap the screen and activate a panic button. When depressed, the operations manager would have to remotely unlock the notepad, but all information contained therein was denied and no transmissions could be made using it. Similarly, the driver's name, location, and deliveries were sent to the regional duty manager and corporate duty manager.

The regional manager confirmed, as part of the plan, that the driver triggered the alarm and attempted to reach the driver to confirm whether an emergency existed. Should the operations manager receive any indication that the driver could be under threat, the operations manager immediately informed local law enforcement, providing them with the access code to the GPS locator in the tablet and all other information. The operations manager then informed the national duty manager. Once local law enforcement was informed, all other drivers were notified of the situation and told to be particularly vigilant and keep an eye out for the other vehicle. If they detected anything suspicious, they were to call dispatch immediately and return to the central station.

The national duty manager then informed management of the situation, while also gathering a communications and customer service group. These groups then put in play the various communications plans while also identifying any similar activities or reports from other regions.

Once the driver and vehicle were recovered, ABC Transport would have a debriefing process involving representatives of management, another driver, and labor. This group would seek to identify what happened that allowed the driver to be put at risk and then would propose potential solutions to the regional management. This measure was designed both to prevent future occurrences and also to attempt to prevent potential labor issues by involving them as early as possible in the hunt for solutions.

Notes

1. Associated Press, "Fog Strands Heathrow Passengers for Fourth Day," *CBC Online*, December 21, 2006, http://www.cbc.ca/world/story/2006/12/22/fog-heathrow.html.
2. These terms of fines imposed and their duration are purely hypothetical in nature but are reflective of possible regulatory implications in real-world scenarios.
3. The term *fog* is used to describe an organization that has no clearly defined goal or direction or one that appears to be misguided or misinformed.
4. Negative value in terms of loss of revenue to the organization or corporation.
5. Translation: This is (essentially) information assurance management that asks the question, Is the information valid, and can I use it?
6. Translation: This answers the question, Are they trustworthy, and can they do the job?

7. *Follow the pipe* is a term used within several industries within the homeland security profession. This approach is a method of indicating that security flaws, leaks, and incidents should be followed to the primary source rather than assuming or taking for granted what has been presented or given; essentially, if in doubt, analyze and investigate where necessary and conclude where appropriate.

8. Under the Safety of Life at Sea (SOLAS) Convention, there have to be enough lifeboats on board to provide evacuation capability for every person. This type of requirement repeats across several similar systems. This results in situations where the vessel might be fully capable of sailing from a mechanical standpoint but cannot sail due to the lack of equipment required to be present. From a disruption point of view, this also provides a possible avenue of attack where, instead of assaulting the hard structures, relatively fragile or separable infrastructure is attacked.

9. This lack of skills often referred to in backpacking and sailing magazines is caused by individuals becoming overconfident in systems. This includes situations where persons in charge of vehicles (in this case aircraft) actually decided to rely more on global positioning system (GPS) technology than on what were previously considered to be safe procedures. The results were avoidable accidents. Ian Mallet, Nick King, and Mike Nendick, "DIY Approaches," *Flight Safety Australia*, November–December 2002, http://www.casa.gov.au/fsa/2002/nov/34-36.pdf.

10. This concept is neither new nor unique. The system is well enshrined in systems such as Carnegie Mellon University's Operationally Critical Threat, Asset, and Vulnerability Assessment (OCTAVE) risk assessment model that describes destruction, modification, and loss (including interruption). It is also enshrined in other systems dealing with threat, risk, and vulnerability assessments outside of the information technology security domain.

11. An *all-hazards approach* involves perceiving threats holistically that commonly occur for any disaster or emergency condition, event, or circumstance, which might include elaborate, but flexible, activation and coordination plans, a communications plans (with contingencies), and general plans that can provide the basis for responding to almost every unexpected condition, event or circumstance. *All-hazards* does not literally mean being prepared for anything and everything; however, any and all plans implemented need to be adaptable, sometimes innovative, and, when necessary, improvisational: Essentially, expect the unexpected.

12. Sometimes referred to in the IT security industry as "zero-day" (or "0-day") occurrences; essentially, this indicates whether a critical feature or function that has been identified as a risk (low, medium, or high) can operate with or without impacting the organization in question. This term is used extensively with cyber-related security systems especially when dealing with, for example, cyber terrorists and hackers.

13. Within the realm of loss prevention, the triad for theft is generally defined by a factor of motivation, opportunity, and rationalization by Dr. Donald Cressey. This involves internal employee theft and should not be confused with factors dealing with whether or not an attacker (e.g., involved in a violent crime) will actually assault the organization.

14. The latter refers to the attack on March 10, 1995, in which a Japanese terrorist group, the Aum Shinrikyo, used Sarin gas as a weapon at five different points in the Tokyo subway system. Some background reference that places this

attack in context can be found at Kyle B. Olson, "Aum Shinirikyo: Once and Future Threat?" *Emerging Infectious Diseases,* Special Issue, http://www.cdc. gov/ncidod/EID/vol5no4/olson.htm.

15. IM/IT is used throughout Canada, where the Canada Information Management/Information Technology (IM/IT) Program helped to strengthen the voluntary sector through technology. It was part of the Government of Canada's Voluntary Sector Initiative (VSI), a joint initiative between the Government of Canada and Canada's voluntary sector. The VSI works to strengthen the voluntary sector's ability to serve Canadians while enhancing its relationship with the Government of Canada. This Canadian program is very similar to the U.S. information-sharing initiatives through Information Sharing Analysis Centers (ISAC).

16. *Risk* can be created by any event or outcome that has the potential to interfere with an agency's ability to achieve its mission; Loss Prevention Risk Finance Section, Office of Financial Management (hereafter referred to as OFM), "Risk Management Division Guidelines for Section 12.1—Risk Management and Self-Insurance Premiums, 2007–09 Budget," *State of Washington,* http://www. ofm.wa.gov/rmd/budget/sec12guide07-09.pdf.

17. "[In security, the] portion of risk remaining after security measures have been applied." Institute for Telecommunication Sciences, "Residual Risk," *Telecom Glossary 2000,* http://www.its.bldrdoc.gov/projects/devglossary/_residual_ risk.html.

18. *Risk appetite* is the amount of risk that an organization is prepared to accept or be exposed to. Risk appetites are often referred to as high, medium, or low, but actually the analysis of risks by an agency includes deciding at what level of outcome the benefit from addressing the risk outweighs the cost of addressing it. In addition, if a program has inherent risks, then having a high-risk appetite and managing the financing of the risk often becomes the risk management strategy. A high-risk appetite is unavoidable for organizations performing inherently dangerous activity or an activity involving the behavior of third parties. OFM (see note 16).

19. The term *public confidence* is in reference to how the public has assurance of the infrastructure, its security, and its reliability regarding moving people and goods to their final destination.

20. Another way of categorizing these options involves options to avoid, reduce, mitigate, transfer, or do nothing. This approach is common in engineering-based systems. These categories, however, assume that management has accepted all arguments associated with the generation of the risk picture. This is not always the case. In this structure, the reduce, mitigate, and avoid options are included within the address category. John Turnbull et al., "Common Methodologies for Risk Assessment and Management," *Royal Academy of Engineering,* http://www.raeng.org.uk/news/publications/list/reports/ Common_Methodologies_for_Risk_Assessment.pdf.

21. Enterprise risk management is the discipline and its associated processes of applying a risk evaluation to each agency activity and outcome, of identifying root causes of unanticipated or unwanted outcomes or potential outcomes, and of determining—as an enterprise—what changes are best to address the root cause and then monitoring the success of the mitigation strategy. OFM (see note 16).

22. Used to legally defer risk away from the parent organization onto a subordinate organization, such as a contracting agency or organization; thereby, that organization that is assuming the risk is the transferee of the risk, and the organization contracting to another organization is the one that transfers risk. However, in some industries, regardless of circumstance, or that have the actual material, such as with transference of nuclear or hazardous materials, the entire chain of custody is legally responsible from initiation to completion.

Local versus Systems Approaches

5

Objective: At the end of this chapter, the reader will be able to define and describe the following:

- The structure of networks
- The difference among the tactical, operational, and strategic layers in the transportation system
- How each of these contributes to assuring the services provided by the transportation system

5.1 Introduction

The transportation system describes a series of interdependent, interconnected nodes and conduits ensuring movement of people and goods. Unlike traditional views of a node as a terminus, nodes act as gateways and transition points within the system. Conduits provide the means of movement and are defined as either fixed or flexible. How nodes and conduits are used is largely defined by direct or indirect demands for transportation sector (including supporting infrastructure) services.

The organization of this system is stratified into tactical, operational, and strategic levels, each one focusing on different aspects of the system to provide functionality. Each level functions as an independent system (facilities, regional areas, and the system in general) and evolves to keep pace with its environment. At the same time, each level is dependent on the performance associated with the other levels within the system in such a way that the entire system operates in balance.

The primary goal of critical infrastructure protection (CIP) is to certify availability of transportation system services by protecting its core operations without unduly affecting its ability to deliver services. Thus, the CIP security manager faces several challenges as he or she is responsible for tactical, operational, and strategic levels.

5.2 Structures of Networks

There appear to be two types of systems present: the exponential system and the scale-free system. These two topographies describe relationships between nodes and conduits based on how they are organized.

In the first topography, the exponential system,[1] nodes within the system are relatively random but are equally connected.[2] Each time a node is added into the system, it spawns a number of conducts that are one less than the total number of nodes. Each node is connected to all other nodes through these conduits. In this type of system, no node assumes a level of prominence or strategic importance. This was the original concept behind the topography of the Internet. Anyone can remove nodes from the system without creating any impacts; however, as one removes a progressively larger number of nodes, the system begins to fragment, finally reaching a critical mass, eventually leading to a collapse or, even worse, total system failure.

The other topography system,[3] the scale-free system,[4] involves establishing pockets of nodes that are connected through central nodes and providing main points of connection to the remainder of the system. In this structure, an individual node is added as either a central node connected back into the system, or as a satellite node connected back to one of the primary nodes. While this type of system is very efficient, it is vulnerable to an attack against any of the primary nodes or the conduits connecting those primary nodes. These structures are mostly spoke-and-hub structures established to deliver services efficiently in competitive environments.

Each system's topography exhibits its own strength and weakness. The exponential system represents an uncharacteristically robust one in which substantial and coordinated efforts are required to successfully disrupt it. This is due to equal interconnectivity of each node. The exponential system, by its very nature, will contain significant levels of waste that create downward pressure on the viability of system organizations. On the other hand, the scale-free system, similar to spoke-and-hub systems, addresses enterprise risk associated with waste inherent in exponential systems, establishing main corridors and nodes that service regional areas. This allows the system to operate at near-optimum levels. The system is not perfect and is vulnerable as key nodes (which are easily identifiable) provide opportunities to attackers to disrupt significant portions of the system in singular or through concentrated efforts.[5]

5.3 The Flux of the Transportation System

The exponential and scale-free system topographies lend themselves to different levels of service program maturity. The exponential system topography naturally positions itself to identify and gradually remove any weaknesses from the system. Nodes and conduits not being used to their capacity are shifted to work at their highest performance levels possible. This gradually evolves into scale-free system topographies where core conduits surviving attrition and evolution inherent to the exponential model. As these conduits and nodes become aligned, the scale-free model gradually emerges, resulting from continually needing to increase efficiencies of the system.

The exponential system is best described looking at a young organization entering a new line of business and establishing a number of locations. The interrelationship between these locations is subject to a high degree of experimentation as the organization, existing in a relatively chaotic state, works through innovation and creativity to establish a core presence within that market and seize as much of that market share as practicably as possible.

This structure operates more within the operational level of the system as the operational level attempts to establish the infrastructure necessary to assure its forecasted or desired performance levels. Even though nodes and conduits may not be attached to each and every node, dynamics associated with establishing need for innovation and creativity within the system remain a significant factor.

The gradual refinement of exponential system topographies—to achieve efficiencies within the system—slowly results in structures that can be shown or reasonably argued to allow the greatest efficiency within the system. The scale-free system, which shows the greatest efficiency of any variety of systems, becomes more apparent within the overall system structure. It must be noted that this process may be driven at the organizational level, creating a situation where the scale-free structure's nodes are being used as the base for new and evolving exponential networks, as new demands are discovered and organizations position themselves in an effort to try to exploit them to their fullest capacity.

This refinement requires additional layers of control to be imposed at all levels. To achieve a level of efficiency within the system, the strategic framework and operational communications of best business practices steadily refine the processes associated with communications and adaptation within the system. Although this system provides a level of predictability and rigidity in the system, it comes at the cost of resiliency within the system.

5.4 Imperatives Driving Network Component Behavior[6]

Driving forces should be examined to understand pressures associated with any flux of the transportation system. As discussed earlier, there are two main forces in the system. The first deals with assuring the level of performance within the system through removing any hindrances to performance. The second involves the installation of protective measures around the infrastructure to assure the availability of the capacity associated with that infrastructure. These forces are in direct opposition to one another, particularly if applied using traditional security regimes.

At the same time, gradual evolution of the system from an exponential (innovation-driven) to a scale-free system (efficiency-building-driven) drives a change in how concepts shift, particularly with respect to threats, risks, safeguards, and vulnerabilities. Though the general concepts remain reasonably valid, their application may undergo a fundamental adjustment.

The first shift involves movement from assurance of performance to gradual assurance of infrastructure. As the number of nodes gradually reduces, the need to protect the key nodes and conduits increases. This is due to the use of a node as a coordination point within the hub as it changes its importance within the system, particularly when compared with its terminus (Table 5.1).

This guarantee of infrastructure drives increasing demand for preventive measures, usually at great cost to the organization. Costs associated with preventing, detecting, responding to, and recovering from security events slowly put increasing pressure on systems to increase their efficiencies of their infrastructure assurance programs as well as their operating costs.

Table 5.1 Performance and Prevention Matrix

	Performance Sacrificed	Performance Sacred
Prevention Sacrificed	Quadrant 1	Quadrant 2
	Focus on the ability to bypass security and impede performance	Focus on the ability to maintain performance but allow controls to be bypassed
	Attackers and attack planners	Use of system as a conduit
Prevention Sacred	Quadrant 3	Quadrant 4
	Focus on the installation of controls, often to the unnecessary detriment of performance	Focus on the ability to assure an acceptable level of performance
	Overzealous regulation leading to collapse of controls under their own weight	For example, CIP managers, operations managers

5.5 Aligning Imperatives with the Mission Statement

Looking at the four quadrants from Table 5.1, various imperatives either align or come into conflict with the overall mission statements.

Quadrant 1 (both prevention and performance at risk): in this environment, the attacker is able to focus on identifying vulnerabilities within the system and then, if possible, on exploiting them to cause disruption within the organization's system. While the system exists in a reasonably chaotic state (exponential-system topography), the focus leans away from attacking any one particular node into one of the following kinds of activities:

- Identification of specific vulnerabilities allowing risk into the system, concealed by similar numbers of potential targets
- Identification of most efficient number of individual nodes that would have to be disrupted to accomplish system-wide disruption
- Reconnaissance attempting to identify potential key nodes and conduits

Quadrant 2 (performance sacred, prevention sacrificed): focuses on establishing abilities at the highest level of performance, even at expense of baseline CIP activities. This dynamic is found during the initial start-up phases of a business line in which a slightly higher level of localized risk is tolerable. During this period, a number of potential activities could be observed, which would include the following:

- Attempts to identify specific vulnerabilities at specific points in the system that could be used to insert something harmful or inappropriate into the transportation system
- Attempts to enter items into the system, often taking advantage of alternate or indirect routes that offer a degree of success
- Attempts to use the system as an avenue of communication, transportation, or facilitation in terms of being able to move items leading to secondary benefits (e.g., narco-terrorism using the system to gain funds)

Quadrant 3 (prevention sacred, performance sacrificed): tends to occur when the system moves too far toward establishing efficiency and controls on the system. There are two reasons for this. The first, and most obvious, involves the strategic level imposing controls to address perceived vulnerabilities in the system (corporate headquarters or regulators). These controls are passed down into the system in terms of constraints or similarly binding rules without due consideration for the impact on performance.

The second reason involves the increased business pressure for efficiency constraining the organization by becoming more focused on process over the final deliverable. This is generally through a misapplication of the best business generation process where the process used to generate a best business process itself takes over the focus, making innovation and creativity a secondary effort. Gradually, this pushes innovation and creativity out of the system as the tactical level (which often generates or initiates the process) sees efforts involved in the process as outweighing benefits.

Controls generated are due to a perceived need to maintain order at strategic levels. Where this process would evolve naturally, as information was passed through the tactical to operational to strategic levels and back down, this condition describes where the strategic level has begun to believe itself to be reactionary and puts stops on the system to regain its ability to control its environment.

The core areas of risk in Quadrant 3 involve those internal to the system, not necessarily generated from outside the system. As the environment becomes increasingly proscriptive, the weights of rules (including regulations) within the system are put under increasing pressure due to the expectations associated with past performances. Thus, pressures evolve and become increasingly apparent within the system:

- Decreased abilities to partner between management levels (loss of communications)
- Increased numbers of components in the system that shift emphasis from working within the system to finding ways to evade or defeat strategic-level priorities (e.g., legal challenges)
- Increased numbers of unreported vulnerabilities in the system and willingness to apply quick fixes to avoid regulatory regimes

Should the transportation system or its components operate in Quadrant 3, the challenge for lead agencies and investigative organizations lies in their ability to quickly gather information and data that would lead to the detection of risk in the system. At the same time, strategic-level controlling organizations need to increase efforts to reestablish trusted communications that may have eroded by conflicts across the system. During this period, an attacker might have an opportunity to investigate the system, to detect any vulnerability, and then to attempt to exploit them with the highest confidence that the controlling entities will either not be informed or will be delayed in receiving any information.

Quadrant 4 (performance sacred, prevention sacred): provides the ideal condition for the transportation system, reflecting a partnership between conflicting values (preventive measures and performance). In this quadrant, the two opposing views attempt to operate in balance—a condition that allows and promotes communication between both systems.

Within this quadrant, an opportunity exists for the CIP manager to shift preventive measures as a brake on the system.[7] This is accomplished by recreating conditions within the exponential system's competitive environment while focusing on a controlled communications and growth process.

5.6 Relationship between Imperatives and Levels

Harmonizing various levels of the organization can be a significant challenge to CIP security managers; this effort involves understanding domains in which systems operate, the operations of the system across management levels, and the shifting priorities within that system. By harnessing various aspects and characteristics of these levels, the CIP security manager can bring about significant value to management within the organization.

5.7 Tactical-Level Imperatives

Maintaining performance within Quadrant 4 is an effort that has institutionalized adaptability and innovation. At this level, the environment is nearly Darwinian in nature as the organization may face significant pressures against its viability should it fail to either perform or protect itself. Local offices must meet performance criteria supported by the organization. Otherwise, the system looks at in similar contexts as a weak node or conduit, paring it away from the overall system to relieve any pressure on the organization. The pressure on the management and organization is first to prevent events and then to adapt to survive those events.

In our continuing examples involving ABC Transport, their warehouse faces a challenge in that it can either deal with criminal events that are occurring at its location or lose a business line supporting the overall organization. Additionally, there are concerns that should those business lines be reduced below acceptable profitability levels, the regional office may decide to close the office entirely before it incurs any further financial loss.

Incidentally, the manager of the local office has examined the company's operations and has determined that most of the burglaries occurred during periods when the office was closed, usually for short period in the evening. The manager also has noticed that the company loads many vehicles during the start of the business hours, which reduces the time it has the trucks arrive when its clients are operational. The manager decides to experiment, shifting the loading schedule to off hours to make sure that four or five people receive security awareness training as to how to report suspicious behavior and contact information should they feel like an incident may develop. The manager hopes that doing this will reduce any opportunities for criminals to burglarize the premise and will increase the number of hours that the company's trucks are on the road making deliveries.

The strength of the tactical level involves working innovatively within a context to force a balance between performance and prevention. The manager cannot simply emphasize additional preventive measures without looking for some way to improve performance or without further trapping himself into a vicious cycle of pressure on measurable performance and increasing the office's risk of being declared a weak node. Similarly, the office manager cannot ignore or accept risks associated with any disruption or losses as they have already surpassed the organization's risk appetite at the operational level.

5.8 Operational-Level Imperatives

The operational level desires to provide abilities to maintain levels of performance within the system, to identify potential issues, and to find methods to mitigate them. However, an additional option exists in this context: a limited option of risk avoidance. This does not exist at the tactical level; operations are demanded by higher levels (meaning that they cannot be simply stopped by local management). The manager may decide to route the organization's operations in such a way that certain environments are avoided.

Returning to our distribution node example that was plagued by criminal problems, the operational manager is currently exploring this option by entertaining notions of relocating the distribution node to another part of the city with a lower level of risk. Given that this option to relocate could bring about significant financial costs and efforts, the operational manager is not likely to exercise this option until he is comfortable that other options are exhausted or until he is directed to do so by upper management.

The operational level provides translation points between strategic and tactical challenges. When communicating from the strategic perspective to the tactical perspective, communications focus on achievable goals in which flexibility lies with making decisions for reaching those goals. Essentially, the operational manager channels the tactical level's energy during the innovation process down a corridor, which increases the likelihood of the tactical level's alternative being acceptable within the risk framework of the overall organization. When communicating from the tactical level to the strategic level, the operational level provides inputs into the strategic level's understanding of the situation, either validating impressions or refining perceptions so that the strategic level is basing its judgments on the soundest advice possible. Should a found option meet any of the requirements, the operational level then marries the two contexts together, communicating them across the operational level (refining the context and solution helps better understand its success within the environment) to the strategic level (indicating that a baseline level of performance has been reestablished).

5.9 Strategic-Level Imperatives

The strategic-level imperatives are opposed to the tactical-level imperatives in that they focus on establishing and communicating goals necessary for meeting organizational objectives.[8] This may involve communicating positive goals, such as meeting expectations or performing at certain levels, or it may mean establishing baselines of performance under which performance would fall outside of the enterprise risk appetite. These might include aspects of loss prevention, accident reduction, and waste reduction. Essentially, the strategic level focuses on the ability to perform and is driven by efficiency.

Regarding ABC Transport, the strategic level's involvement focuses on communicating thresholds to the operational level. In this context, the criminal activity is pushing the level of losses above the baseline of acceptable losses within the organization, putting pressure on the organization to resolve the issue. If the issue is not resolved, the strategic level provides the operational level the authority to detach the node from the system, either halting the operations of the enterprise within that area or relocating them to better environments.

The strategic level defines parameters of acceptable options for the operational and tactical levels. This provides a target at which the tactical level is aiming when seeking to meet the challenge of addressing the risk from an innovation point of view. It also provides a context whereby the operational level can guide the tactical level within the innovation process. The strategic level provides validation of whatever solution was determined through an organizational acceptance, or it provides further refinement of that innovation and communicating into other regional organizations as a capacity-building measure.

5.10 Aligning the Levels of the Organization

Establishing this process requires that the organization is working within the context of Quadrant 4 across all three management levels. If the strategic level focuses on Quadrant 3, it will more likely prescribe the solution, thus losing an advantage of the innovation inherent at the tactical level, but will place the organization at risk if it is associated with institutionalizing a decision without adequate situational awareness.

At the tactical level, Quadrant 3 is less likely to be applied, except in local management decisions, as there are no delegated levels closer to the work. The emphasis on Quadrant 3 at the strategic and operational levels, however, could force the tactical level into making decisions based on Quadrant 2. This is due to attempting to reestablish favorable performance levels (a measure of self-protection) but could also lead to friction among tactical, operational, and strategic levels that are more in line with a backlash against their adoption of Quadrant 3's position. This generally takes the form of demands for more resources and complaints against the operational and strategic levels' unwillingness to address concerns of the tactical level. This scenario can quickly cascade within and into the operating environment, promoting a number of challenges. These range from lack of awareness (tailored communications informing the operational level and strategic level that the "situation is under control" when it is not), falsified reporting of performance levels (through either misleading or not reporting), or even the increased risk of operational disruptions (e.g., labor-relations issues).

If strategic or operational levels begin to focus on Quadrant 2 (performance sacred, prevention sacrificed), decisions will cascade down through the delegation mechanism to tactical levels and possibly be imbedded within the organizational culture at all levels of the organization. This opens the organization to the risk of creating conditions within a regulatory community to impose conditions of Quadrant 3 on the enterprise system. This is due to the perception that the organizations making up the remainder of the system are not adequately motivated to meet certain requirements. If this condition is reversed (tactical adopting a focus on Quadrant 2 and ignoring requirements of operational and strategic levels), the organization faces internal discipline issues that would be corrected through application of its own measures to avoid outside interference within its internal affairs.

When the organization and its levels begin to operate in Quadrant 2, the perception of greater rewards are achievable and sanctions are associated with factors of performance. If the costs associated with a fine are $200 with no further repercussions and possible incomes associated with that violation are $2,000, a condition exists where the management could adopt to absorb the costs associated with any fines incurred as part of the cost of doing business. On the other hand, a decision to focus within Quadrant 3 generally stems from a condition within the organization that leads to a condition of mistrust or similar factors that compels the operational and strategic levels to become increasingly involved at the tactical level.

5.11 Communications among the Levels

Communicating among levels requires that each level perform some validation of its decision-making capability to minimize risks associated with these shifts. The strategic level must determine whether or not its goals are achievable, delegating to the operational and tactical levels. Similarly, the operational level requires validation of these decisions (generally reporting the decisions suffice) but also feedback regarding specific conditions at its tactical level. Finally, the tactical level requires assurances that its senior management supports its decisions, given that resources generally are assigned from strategic to tactical levels.

5.12 Pace of Evolution

Interaction forces various levels of organizations within the transportation system to evolve at different rates. For the tactical level, evolution based on reaction to events is swift. On the other hand, if management allows a decision to be ensconced within an organization, it may be delayed for some time while the operational and strategic levels both take their time to review the implications associated with that decision before communicating it across the organization. This can put pressure on the tactical level to begin to shift back toward Quadrant 2 in the system unless the operational and strategic levels respond to the tactical levels' communication and establish some level of dialogue to lead toward a solution.

At the strategic level, a number of processes and considerations naturally slow the evolutionary process. The implications of information received from the tactical level have to be viewed through a strategic lens. This means that tactical-level factors, if not first identified at the operational level, must be identified and then "tagged" as evaluation points. These evaluation points become important in the lessons learned and best business practices activities of an organization as they assist in determining whether the information provided is both valid and useful. Once tactical- and operational-level data within the communication are assessed, the strategic level must then decide how to approach issues of incorporating it within the strategic system or of identifying it as an option. This is done to maintain the direction of the organization while monitoring and maintaining levels of control over evolutionary processes. Management may continue planning and charting activities for medium- and long-term goals.

At the operational level, the pace of evolution is dictated by interaction between the demands of the strategic level and conditions at the tactical level. This invariably leads to a level of tension within the system. At one end of the spectrum, the tactical level will continuously seek guidance and approval for its reactions and actions to make certain that its resources and its actions promote its standing and well-being within the organization. The nature of the strategic level, with its requirement to judiciously study organizational impacts, results in a slower approach of management. This tension is one of the foundational reasons for ensuring that operational managers are delegated a level of authority that is clearly defined so that an executive or manager can make necessary business decisions without putting mechanisms at risk.

5.13 Internal Influences versus External Influences

Evolution has generally been perceived in terms of the organization growing as a result of external influences. Although evolutionary influences are true to some extent, dynamics change somewhat when looking at the transportation system as a collection of interdependent, interrelated systems.

Internal influences, as the name implies, stem from factors arising from within the organization. Consider, for example, the strategic level's decision to cut the organization's operating budget by 5%. This decision cascades downward into and throughout both the operational and tactical levels, sometimes varying in such a way that the financial risk associated with the performance of certain measures in processes changes. Because of these changes, the organization is forced to adapt into the new environment.

This is an issue in enterprise risk management. Consider, as outlined within the "follow-the-pipe" concept, various inputs to processes. Those inputs support the overall ability of the organization to deliver its final services or goods, as well as within the context of an all-hazards environment. Any risk that has the ability to jeopardize the organization's ability to deliver falls into this category and has to be viewed from a CIP security management perspective. An approach to dealing with this specific challenge is discussed later in this volume.

External influences fall into another series of categories. On one hand, natural disasters and similar kinds of environmental events (e.g., wind shear, storms, current shifts, volcanoes) pose risk to the system, but only through their presence. On the other hand, other individuals or organizations may pose a risk against the organization in terms of their presence, illustrating a level of intent to cause disruption to the organization. This is particularly true in competitive industries where two (or more) competing organizations are attempting to seize control of a larger, but common, share of the market.

5.14 Transorganizational Constraints

The transportation system is composed of more than organizations simply operating at tactical, operational, and strategic levels. Some organizations operate at levels that are one layer higher than the strategic level for private-sector organizations; these include regulators, law enforcement, labor boards, as well as industry-wide associations or consortiums.

5.15 Alignment with Mission Statements

These organizations tend not to represent a specific business activity within one organization but to influence a kind or range of activities within the transportation system. Regulators, for example, tend to place constraints on a single kind of industry or activity, attempting to address issues that arrive in that specific field and usually dealing with compliance issues of industry-imposed rules or laws imposed by countries. Other groups tend to work toward representing a single kind of activity across the system or at least within an area they have the power to influence. Labor organizations, for example, may represent either directly, or as a coalition, individuals who provide a specific set of services within the system.

5.16 Influences on Follow the Pipe

These organizations operate slightly differently within the system as they influence a type of activity that may be present in many different systems when examining various pipes. This means that a decision to exert influence at a tactical, operational, or strategic level does not exert influence in one specific system but across the entire geographic area, which defines the influence of that level of the organization and its control.

Consider, for example, British Columbia's trucking labor strike at several maritime ports during July 2006. The main trigger of this dispute involved a sudden rise in the cost of fuel that caused many truck operators to operate at a loss. The impact of the strike crossed several organizational lines. While the ports were obviously impacted, other industries were also affected as they were unable to move their goods either into or out of the port areas.

In 1993, a similar dispute at one of the West Coast ports in the United States had a similar impact. During that dispute, the inability to move grain through the port system had impacts all the way back to the U.S. heartland, which saw its ability to harvest affected due to the inability to free up space because of the blockage in one part of the system.

5.17 Alignment of Transorganizational Groups with the Matrix

Transorganizational groups face similar dynamics as a private-sector organization when looking at their own missions within the transportation system. For example, a labor organization that chooses to disrupt the overall system would probably fall into Quadrant 1—although not to the same degree as a determined attacker who seeks to cause permanent damage to the system. The difference is that the labor disruption is used in the system to make a point regarding the value of the service through revealing the impact associated with its removal, and there is intent at some point to resolve the issue and return the system back to an operating level. This consideration does not factor into a terrorist attack.

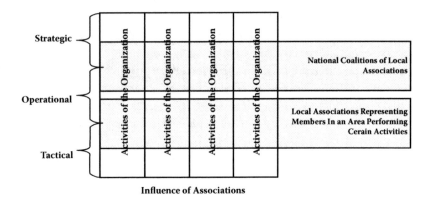

Figure 5.1 Participants and influence in the system.

The labor organization can also fall prey to Quadrant 2 and Quadrant 3 issues with respect to its mission of conducting negotiations. Although this does not necessarily align completely with the concept of performance and prevention, if a manager makes a subtle distinction between maintaining the services provided by its membership (performance) and hard-line negotiation (prevention), he or she will find similar kinds of dynamics. Where the labor association falls into Quadrant 2, the tendency would be for the reporting chain within the organization to be eroded, with complaints or issues being held at the local or regional levels. In Quadrant 3, the hard-line approach could cause a push for the system to look for replacement workers (nonunionized personnel) as management attempts to escape the constraints of the collective agreement.

5.18 Constraints by Regulators

The regulator does not tend to operate within Quadrant 1, as its task is to maintain appropriate performances of the transportation system, or at least particular activities within the system. Where the regulator tends to operate is between Quadrants 2 and 3, often moving between the two on an issue-by-issue basis. For example, the regulator may attempt to focus exclusively on maintaining operations of a transportation system to avoid conflict within the system but would put into place systems allowing for certain kinds of controls, checks, or balances to be bypassed. The regulator can choose to emphasize the control aspect of the system, running the risk of creating cumbersome and burdensome regulatory regimes and putting severe strains on industries regulated. These constraints put into place by regulators tend to operate across an industry, often capturing several activities and operations within the regulatory net. For example, the Marine Transportation Security Regulations in Canada, the various rules of the U.S. Coast Guard,

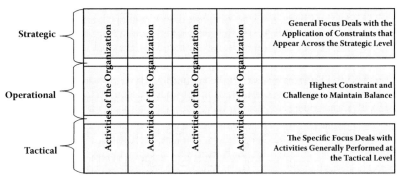

Figure 5.2 Focus of constraints.

and the regulations associated with civil aviation. The sum of these various restrictions on their activities is intended to operate as a system that raises the baseline of security within the overall mode, industry, or activity.

Constraints issued by the regulator tend to affect the tactical and strategic levels in an organization the most. Organizations are required to put into place systems or capabilities across the entire organization. This may include aspects of prevention, detection, response, recovery, and reporting.[9] At the tactical level, the location may be subject to observation and inspection of its processes and procedures to determine whether specific constraints have been put into place. Similarly, the strategic level, once it has passed its own requirements and adapted its system, may shift the impact onto tactical and operational levels, indicating that any sanctions, such as fines, will be drawn from the operational or tactical level resources to minimize the impact on the corporate bottom line and to provide negative incentives for noncompliance.

For the regulator the indicators associated with its operations in either Quadrant 2 or Quadrant 3 would generally arise from industry groups and other associations that are aligned because they all fall under the regulations. These issues may be discovered at tactical, operational, and strategic levels of interaction, requiring the regulator to maintain lines of communication across all levels to identify and, if appropriate or deemed necessary, to respond to those issues (Figure 5.2).

5.19 Questions

1. Identify at which level an activity is most likely to be found using this model:

   a. Installing a CCTV system
   b. Identifying alternate sites

    c. Hiring a guard force

    d. Providing the methodology for how risk is going to be communicated

2. How would you describe the four possible alternatives when looking at attempting to balance performance and protection in the system?

3. The manager of ABC's facility has identified a serious internal theft issue, and the decision is made to implement a background screening system for all employees. Briefly describe the role of each level within the organization.

4. Why is it important to understand the relationship between nodes and conduits in the transportation system?

5.20 Answers

1. (a) tactical; (b) operational; (c) tactical; (d) strategic
2. This involves the combinations of performance and protection being considered either sacrificed or sacred.
3. Strategic—setting the overall policies requiring a background screening; operational—describing the factors that could influence the results; tactical—taking of applications, feedback on the specific roles and location.
4. The nodes and conduits will have an influence on the impact throughout the system. For example, this might include disrupting a conduit between two well-protected nodes.

Notes

1. These two structures are described in Reka Albert, Jeong Hawoong, and Albert-László Barabasi, "Error and Attack Tolerance of Complex Networks," *Nature*, Vol. 406, 2000, 378–382.

2. Although many of the arguments defined within that work would pertain here, because the transportation system is constantly evolving and shifting means that it involves a hybrid of the two systems with the characteristics closely tied to the business maturity, infrastructure challenges, and activities associated with the activities taking place between the various nodes and conduits.

3. A *scale-free network* is a noteworthy kind of complex network because many real-world networks fall into this category. *Real world* refers to any of various observable phenomena that exhibit network theoretic characteristics (see, e.g., social network, computer network, neural network, epidemiology). In scale-free networks, some nodes act as "highly connected hubs" (high degree), although most nodes are of low degree. The structure and dynamics of scale-free networks are independent of the system's size N (the number of nodes the system has). In other words, a network that is scale-free will have the same properties no matter what the number of its nodes is. "Scale-Free Network," *Wikipedia*, http://en.wikipedia.org/wiki/Scale-free_network.

4. John Robb, "Scale-Free Networks," *Global Guerrillas,* May 7, 2004, http:// globalguerrillas.typepad.com/globalguerrillas/2004/05/scalefree_terro.html.
5. This can operate on two levels in the second model. The first involves a disruption of a node. One cannot discount the consequences associated with the disruption of a key conduit that connects these primary nodes.
6. C. S. Holling, "Understanding the Complexity of Economic, Ecological, and Social Systems," *Ecosystems,* Vol. 4, no. 5, August 2001, 390–405 provides a good overview of this kind of approach. Once again, however, the level of maturity and evolution in the various systems would not necessarily be identical but would be driven independently and then either accelerated or decelerated according to the interdependencies and interrelationships with other parts of the system.
7. Preventive measures are measures or countermeasures representing checks, balances, or even business enablers within any given system.
8. This is often remarked upon as being a challenge, both by the Office of the Auditor General in Canada and the U.S. Government Accountability Office. Though both offices tend to look at government activities and performance (as regulators), the principle is the same. Consider, for example, the role being played by the U.S. Federal Emergency Management Agency (FEMA) in the Emergency Alert System described in GAO-07-411.
9. The concepts of prevention, detection, response, and recovery parallel other systems, including those put forward by such professional associations as the American Society for Industrial Security (ASIS), which espouses the concepts of deter, delay, detect, deny, and detain. The end goal is to be able to put into place appropriate measures that allow an entity the greatest opportunity possible to halt an attacker before he or she may reach the end goal.

Criticality, Impact, Consequence, and Internal and External Distributed Risk

6

Objective: At the end of this chapter, the reader will be able to do the following:

- Identify risks within, throughout, and external to the organization and how risk interrelates with the organization's goals and practices.
- Identify levels of criticality and methods of impact, and their consequences as to how criticality levels of measured reliability and accurate impacts are reviewed.

6.1 Introduction

The factors of criticality, impact, and consequence play pivotal roles in assuring the availability of the transportation system. Understanding these concepts, their interrelations and how they interact with the infrastructure are extremely important for the critical infrastructure protection (CIP) professional attempting to identify risk in the system.

The transportation system represents the sum of a number of parts working toward relatively aligned goals. On one hand, private-sector organizations fulfill market demands through the provision of goods and services. These efforts are supported by government operations that provide key components of the system, such as navigation systems. Governments also tend to set limits on the operations of systems, particularly where there may be a perceived risk to society. Each of these components of the system has its own infrastructure supporting these operations.

For the CIP security professional, criticality, impact, consequence, and risk can operate at any combination or permutation of the tactical, operational, or strategic levels. To ensure that the system is still capable of performance, professionals must examine and understand demands being met at all of these levels.

It should be noted that CIP exists beyond the United States and Canada insofar that discussion is internationally based. Financial representation is non-denominationally defined as a *transdollar*, defining no one particular country.

6.2 Assignment of Value

Taking the paraphrase from the mission of ABC Transport—"moving persons and goods to the right location in the system within the right time frame and in the right condition"—several systems can be used to measure the performance of the system. For example, consider the number of individuals who were redirected, delayed, or injured for any trip. In the context of goods (e.g., packages, cargo), this might also refer to the number of goods lost, redirected, delayed, or damaged.

Thus, we need to consider the two general missions of the CIP security professional. The first goal is the maintenance of the system's capacity to achieve that mission. This involves ensuring that individuals or goods are able to make their transit successfully. The second goal involves the protection of the actual system from attack or inappropriate use. The attack concept depends on the assurance of infrastructure necessary to support the first goal. Second, the government may be imposing social restrictions. The CIP security professional has to remember that these considerations are meant to be presented within the context of an all-hazards approach, not from the narrow confines of one type of attack or loss within the system.

The performance of the organization is directly related to the availability of resources within the organization. These resources are necessary to provide the inputs that complete the processes capable of achieving those missions of the organization. Given that management provides the resources at any of the tactical, operational, or strategic levels, managers can begin mapping the importance of assets back through the management consideration involved in allocating those resources. This might be described in terms of Figure 6.1.

Above the forecasted levels, management stresses are reasonably positive, and the organization does not face a significant risk of running out of resources. In essence, the organization's performance is resulting in the necessary profits that allow it to grow and evolve in accordance with its management plan. This may be considered a positive environment.

Between the forecasted level and break-even point (BEP), the organization continues gathering profits to meet planning requirements, though not at a rate supporting the management plan. Management priorities may focus on improving efficiencies of the organization, but not through drastic actions.

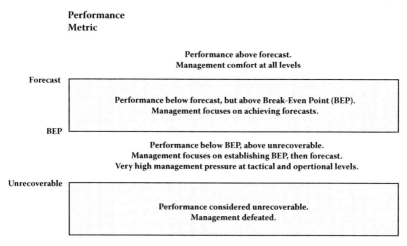

Figure 6.1 Sample management scale for "criticality."

Between the BEP and unrecoverable levels, the stressor on management involves reestablishing the necessary levels of gathering profits. At this point, the organization is depleting its ability to generate additional short-term resources (e.g., through credit) and reserves. The management may consider casting away efforts that are a particular drain on the organization or may take measures to identify and reduce inefficiencies in the system.

Finally, at the unrecoverable level, the management of the organization decides to close down the organization.

6.3 Criticality

Criticality looks at the contribution of an individual, asset, facility, information, or activity to the success of a process and, through that process, the performance of the overall system. Removing or changing the contribution of any one of these inputs may result in an immediate change in the potential to reach the intended level of performance in a process. The magnitude of this immediate change (which operates independently of time) is referred to here as the criticality of the input.

It is unlikely that persons, assets, facilities, information, or activities are unique to only one process. In this context, one also has to look at the overall contribution.

6.3.1 Single Points of Failure

Single points of failure (SPoF)—in any system—are reasonably undesirable and can operate across one or more processes. These conditions exist where there are no suitable replacements for the contribution being made by a specific point in the process such that loss of that contribution halts processes from continuing. This drives criticality of SPoF to a level that reflects value of the system in which it operates.

Single points of failure may be the result of a number of conditions. First, the market may only be able to bear one of that particular kind of enterprise in the area due to a relatively limited demand. Second, the service may be mandated to be performed by a specific organization or entity, such as traffic control systems, navigation systems, or specialized communications systems. Finally, if the system is complex, there may only be a very limited number of specialized entities that can deliver the service.

6.3.2 Consideration for Nationally Declared Critical Infrastructure

Whereas *criticality* describes value in terms of process, *critical infrastructure* refers to infrastructure used to support the safety, security, and economic well-being of the citizens of an economy by making certain that the nation is able to continue to function.[1] This definition may change, but the intent remains generally the same. In these cases, government through legislation or regulation generally dictates the value of the infrastructure's criticality.

As with other infrastructure, this can involve individuals, assets, facilities, information, or activities, and as a result, an enterprise may deliver something considered to be a critical infrastructure as either its primary business line or as a secondary business line or subprocess in its entire system. It is important that CIP security managers contact the local emergency management office (FEMA[2] for the United States; PSEPC[3] for Canada) to identify whether or not this condition applies.

6.4 Impact

Impact measures the loss over a period of time. Time becomes important when looking at the three ranges above the unrecoverable level. Should the organization operate below BEP, time influences the full extent of loss. Similarly, if the organization operates above the BEP but below forecasted levels, time will factor into the full extent that the organization's plans may be disrupted. Where the organization is operating above forecasted levels time factors positively in terms of either reversing past losses or increasing reserves associated with future losses.

Within the context of business continuity planning, the maximum allowable down time determines the organization's tolerance for a level of disruption. The tolerance is defined in terms of when the organization has reached a substantial threshold in failing to meet its mission. Though not necessarily always the case, this can generally be attributed to performance-based loss.

6.4.1 Tactical-Level Impact

Impact, at a tactical level, refers to the ability of the node or conduit first to deliver a service over a period of time and then its survivability in the longer term. Tactical impact can be driven through the loss of resources (or the loss of ability to generate more) or, in the case of larger organizations, when the operational and strategic levels of the organization are no longer willing to sustain a particular level of loss any further. For the CIP security professional, it is important to note the impact both on the ability to deliver services and on operational and strategic levels of management.

6.4.2 Operational-Level Impact

Changes in performance at tactical levels define impacts at operational levels. As the subsystem of nodes and conduits are more or less able to handle demand placed on them, for operational levels two questions exist:

1. How much do tactical level entities contribute to performances of systems at operational levels?
2. Can performance be made up by using other nodes and conduits in the subsystem so that the impact is reduced as much as possible?

As the performance of a node or conduit changes, its ability to handle demand also changes. Part of this is due to the natural desire of enterprises to work at a level that maximizes profit or capacity. Just as the loss of transactions due to a security event puts pressure on the business, so does the loss of business that results in a loss in the number of transactions. This could cause a number of impacts and depends on whether or not the organizations within the environment are working under capacity, at capacity, or over capacity. These impacts can be described in terms of the following.

6.4.2.1 Slack and Overflow

The term *slack* describes conditions in which components in the system are able to absorb additional capacity. Overflow, on the other hand, describes conditions in which demands on components have exceeded their capacity to deliver. This can result in partially completed transactions (e.g., storing a container for a period of time) or the shifting of demand into another part of the system.

6.4.2.2 Pull and Push

The term *pull* describes conditions in which entities are not working at full capacity and attempt to attract business to reach capacity. This is accomplished

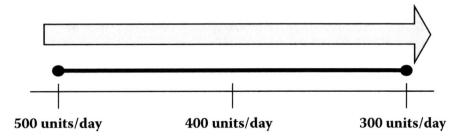

500 units/day   400 units/day   300 units/day

Figure 6.2 Capacity issues between nodes–linear.

by either creating a center of attention for new business into that component of the system or, where two components attempt to deliver the same good or service in that organization, consolidating two components for sake of efficiency.

Push, on the other hand, describes conditions in which components in the system are beginning to place too much demand on another part of the system. This generally results in components looking for new ways to have their demands met (Figure 6.2).

6.4.2.3 *Delay and Lag*

When a component in the system cannot meet the demands being placed on it, there is a risk of *delays* in the system. This delay can be either the result of having to shift to less efficient or less timely routes or of something being caught in the system. Consider, for example, air travelers stuck traveling to a city where the airport has been closed due to weather issues. Flights being canceled results in a delay in the system for those travelers.

Lag results when the component in the system cannot generate adequate demand for another component in the system. As a result, the receiving component attempts to generate business from other sources to operate at capacity (Figure 6.3).

6.4.3 Strategic-Level Impact

At the strategic level, several different kinds of impact can occur. Within the private sector, if a sufficient number of operational levels are failing to meet goals, the impact can spread into strategic levels. This forces strategic levels of management to make adjustments that cascade back to operational and tactical levels. For example, to reduce costs, strategic management demands its operational levels return certain amounts to reduce costs. This demand is then passed on in turn to tactical levels.

At the same time, strategic levels may put pressure on the organizations responsible for constraints in the system (e.g., regulators). These pressures might take a number of forms but generally focus on abilities of organizations

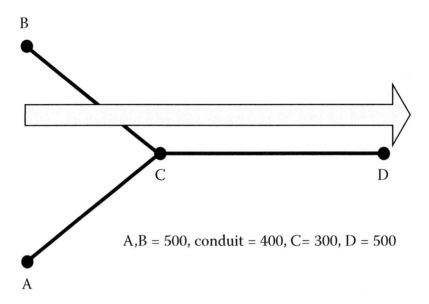

A,B = 500, conduit = 400, C= 300, D = 500

Figure 6.3 Capacity issues—small system.

to protect their own resources. The return pressure might be a perception in the regulatory organizations that industry is unwilling to comply with regulations, leading, under the right conditions, to situations in which enforcement blitzes and other activities to achieve forced compliance take place.

6.4.4 Consideration for Control Systems

Control systems may operate outside of various modes of regulatory controls but may have devastating impacts depending on their deployment. These control systems are involved in managing movement within conduits between nodes in the system and have a direct impact on the rate at which movement proceeds through those conduits. When they fail, these systems lead to accidents. For example, in India in 1999, more than 250 persons lost their lives when two trains collided; this tragic accident has been largely attributed to a failure in signaling equipment. Often these systems are coordinated through centralized computer systems or networks of computerized systems that track both conditions within the system and its status of control points.

6.5 Consequence

Up to this point, impacts that have been considered are internal to the organization's operations. This is not always the sole consideration. Events often lead to obligations to communities that are outside the control of the organization An example might occur when an accident that vents toxic gases may require an area to be evacuated, and the organization, if due diligence cannot be demonstrated, may be liable for costs associated with that evacuation as well as any civil proceedings arising from it. Explosions may damage surrounding areas, resulting in the organization being held liable for costs associated with reconstruction, restoration, and damages within the realm of civil liability. Legal and civil liability, social responsibility, and other similar factors are often consequences to these types of events.

Consequence, due to its outside origin, may be applied at different levels of the organization depending on legislative and regulatory actions in force. Vicarious liability, enterprise liability, and similar legal mechanisms (ranging from administrative monetary penalties through imprisonment for senior officials) all have impacts on the organization but are set from outside the influence of the organization itself.

6.6 Risk

Risk may be defined in terms of factors of probability and impact. Probability might be defined in two terms.

- The first part is based on historical factors, which determine whether something has happened and perform checks in determining if these factors exist that would allow it to happen again.
- The second part looks at forward-looking systems, essentially the likelihood of an event occurring. In this concept, an organization might ask whether the means, opportunity, and intent are within the realm of reasonability for making an attack conceivable.

For this context, impact will incorporate consequence as the combination will eventually be incorporated into the loss picture for the organization. Though this may be the eventual outcome when looking at impact and consequence from an enterprise perspective, it is important to recognize the differences between the two and the spans of control associated with each.

6.6.1 Internal Risk

When significant risks reside within organizations but they are of little consequence, they may be termed *internal risks*, which are an organization's management challenge.

The key exception to this involves the provision of a critical infrastructure in the sense of goods and services that support declared infrastructures as a business or product line. For this reason, goods and services provided (by their nature) have a consequence, meaning that disruptions in the expected service or delivery of the goods can be considered to automatically have an external consequence.

6.6.2 External Risk

On the other hand, impact and consequence both affect the same organization, putting downward pressure on its performance. Special sources of funds are not used to pay claims for damages. When insurance would be seen to cover all of these costs, it is simply an exchange of risks with the main effect being a sustainable and predictable financial risk (insurance premiums) replacing the risks associated with having to pay massive legal fees and penalties. Thus, these risks are termed *external risks*.

6.7 Risk Calculations

The key to calculating risk is to ensure that one method is used consistently across the entire organization. This will certify that tactical, operational, and strategic levels of an organization can communicate information based on common definitions and processes, something that is necessary to being able to collate and compare information.

Two general systems are used to calculate risk. The first involves a quantitative system that identifies and enters values into a formula that results in a score. This could involve a system in which the specific values are input and the final score represents a financial value or an effect on performance. The second system involves qualitative assessments of risk, often used by assigning ranges.

Generally, as probability increases, risk will also increase. The challenge involves balancing conditions in which exceptionally high impacts are associated with exceptionally low probabilities. In today's climate, the dominant factor still resides with the impact half of the balance. This may change as organizations seek to become more efficient and as a return to foreseeable events is desired. Looking back at approaches prevalent within the security industry, there is a swing between the two approaches that might be associated with major events (e.g., the September 11 attacks) and society's reaction to those events.

It is during these periods of fluctuation that intelligence organizations (and other similar organizations) show their true value by being able to calculate likelihood reasonably accurately. When looking at likelihood, the question revolves around means, opportunity, and intent. Essentially, when looking at this, this balances whether or not there is a value to the attack (e.g., ability to disrupt, cause terror, exploit systems) coupled with the presence of the necessary knowledge, skills, and abilities necessary to exploit vulnerabilities in the system.

6.8 ABC Transport Example

ABC Transport operates with an administrative structure of five regions, each of which has a head office and five local offices. Each regional head office also serves as a central distribution node that services the local offices.

Clients can request shipment of a parcel by going to the local offices, making a request by telephone for pickup, setting up a recurring shipment, or requesting a special pickup through the Internet. Approximately 40% of the shipments are set up through the Internet, whereas recurring pickups make up another 30% of the shipments. Telephone and drop-in (will call) shipments account for the remaining 30% with each making up, on average, 15%. All activities are tied to a shipping planning system (SPS)[4] that coordinates the movement of all parcels within the system and then provides information to the client with respect to their location.

ABC Transport is in a very competitive market, leading to a situation in which the profit margin per unit has been reduced to remain competitive. Additionally, ABC Transport has started to implement a plan for expansion onto the West Coast. The financial and operations organizations at ABC Transport compiled Table 6.1 for its managers to clarify the various limits on performance within the organization. Management across the organization was informed that if their organizations did not meet the expectations listed on this table, senior management would likely take actions with respect to the offices failing to meet their performance quotas. This table was based on each parcel costing US$22, of which $2 was profit not allocated to any activities. The breakdown of the price per parcel was the following:

Operating Costs: $18.00
Expansion Plan: $2.00
Profit: $2.00

In Table 6.2, ABC Transport has set up thresholds used as a measuring stick to assess criticality. The first step is to determine the acceptable goal, in

Table 6.1 Sample Table for Goal Setting

	Organization Goal Strategic	Regional Offices Operational	Local Offices Tactical
Cost of Forecast	22,000,000	4,400,000	880,000
Number of Forecast	1,000,000	200,000	40,000
Cost of BEP	18,000,000	3,600,000	720,000
Number of BEP	818,182	163,636	32,727
Cost of Unviable	9,000,000	1,800,000	360,000
Number of Unviable	409,091	81,818	16,364

Table 6.2   Dropoff Levels—Contribution to Criticality

Allowable Drop	Organization Goal Strategic (%)	Regional Offices Operational (%)	Local Offices Tactical (%)
Forecasted	0	0	0
Break-Even Point	18	18	18
Unviable	59	59	59

this case the forecasted goal. For this level, no drop in performance is below the forecasted toleration level. The second threshold becomes the division between profitability and loss, and in this case anything higher than an 18% reduction in performance is assessed to lead to a condition of loss. In this case, ABC Transport's management has decided that a loss of 59% performance across the board will signal that the organization is no longer viable.

The values associated with the cut-off points in Table 6.2 are used when looking at criticality during the follow-the-pipe exercise. Again, criticality operates independently of time; the impact is based on the length of time the problem persists. For the purposes of giving an example, consider the Web-based booking system for clients.

The loss of the Web-based booking system leads to a reduction of new items by 40% per unit of time, as one looks at the historical statistics. Even though this number may or may not be completely accurate due to small changes in the system, it does provide a reasonable basis on which to make a prediction.

This 40% reduction leads to a drop in performance between the BEP and the unviable range. When approaching this issue, the priority of work is determined by examining the extent of the impact and consequence apparent in the system. In this case, the level of disruption makes it the second highest category of criticality (referred here as a Priority 2 condition). This Priority 2 designation means that in the case of disruption, it would have less loss per unit of time than a Priority 1 asset but more than Priority 3 and 4 assets. For management, this means that where disruptions were occurring they would prioritize the work from Priority 1 to Priority 4.

The next step looks at the impact on the organization. If the Web-based booking system goes down or offline, there are two impacts. The internal impact is that the system loses approximately 1,096 parcels per day coming into the system. This would force the overall organization below the break-even point in approximately 166 days. Externally, one consequence would involve public confidence in terms of not being able to enter a parcel. If they cannot maintain a booking system, why should they be trusted to make deliveries? What else is wrong here?

This impact carries on down the pipe and involves the connection between the Web-based booking system and the SPS. The loss of this connection means that the two systems would no longer communicate. For the booking of parcels, this is similar to the lack of the Web-based system itself, although the data flow will factor heavily in this respect. On the other hand, clients will not be able to check on their package's arrival, meaning that they will not pick it up unless contacted directly by ABC Transport. A manual notification of clients would be awkward at best.

Moving further along the pipe the impact comes to the SPS that is tied to all planning activities and booking systems and that provides a core service. Consider that it can be assumed that 85% of the booking system would be directly affected as the booking agents cannot schedule a pickup of the parcel. For those that are being dropped off, the rate would slow somewhat, but a manual system might be used to collect the package.

In this case, the disruption of 85% of the incoming parcels is well below the 59% threshold that the company has set for viability. Thus, the SPS is a Priority 1 critical system and will receive immediate and unwavering attention until its services are restored back to expected levels, even if it means pulling people away from a Priority 2 task. The impact associated with this system is such that the company would be declared unviable if this was the only loss and if it was allowed to persist for 175 days. The BEP of the organization would be reached in 78 days; the company would see this as a grave situation.

If these systems were to fail locally (not nationally) a different scenario would exist. In this case, local performance is disrupted. As a result, numbers associated with the local offices would be used to determine criticality and impact, not the national numbers. At the same time, close observation would be wise to determine when the local office's performance began to have a significant impact on the regional office. This might be considered to be at the point where the losses at the local office began to draw resources out of the regional office, forcing the regional office to draw excess performance from other parts of the system.

At the same time, there is a lag in the system created in this situation because the local office is not performing to capacity. The level of performance of the local office is not meeting the demands of the regional office, and if the loss is allowed to persist for some time the regional office may attempt to draw additional business from other sources. When the local office comes back on line it may experience a level of delay if the regional office is suddenly forced to work beyond capacity.

In the case of ABC Transport, all nodes and conduits are performing equally, meaning that the loss of any one node or conduit would have the same impact. If half of ABC Transport's business was conducted between two cities this would shift the dynamics somewhat. Similarly, if the disruption were to happen just before Christmas or some other high-demand period, another shift would happen.

This kind of loss could also result from situations beyond the control of ABC Transport. Significant weather events could have an impact on a region, such as seasonal hurricanes in the United States. Government regulators could shift the location of the BEP and that of the unviable threshold by creating regulations that impose new costs on the organization. An addition of a surtax of $1 per parcel, for example, could have a significant impact on the competitiveness of ABC Transport or could cut its anticipated profits in half.

Whereas the organization is focused inward on its performance and its ability to maintain market share, the system is preoccupied with the potential loss of capacity within the system. ABC Transport deals with what can essentially be a special delivery service, moving nearly one million parcels per year. Given the nature of the movement and the volume, what is the effect on the overall system? Does ABC Transport provide a critical service reflected in emergency plans (e.g., the shipment of volumes of centrally stored vaccines to hospitals in the case of a pandemic)? What does this loss of capacity mean to the capacity of the entire system? Can the loss of ABC Transport, for example, simply be written off as a business loss as the other competitors will move in quickly to pick up the slack? Having identified these risks, the next step is to look at options on how to mitigate those risks.

6.9 Questions

Risk may be viewed in terms of impact and _____.

- Probability. Often linked to likelihood, probability involves the chances of a particular outcome occurring within a set number of attempts. In this context, the presence of vulnerabilities that can be exploited by an attacker possessing the necessary knowledge, skills, abilities, and resources would raise the probability of a negative event.

The most important factor of impact is (a) time or period of time of an event; (b) correlation of an event; (c) probability of an event; or (d) consequence resulting from an event.

- (Answer) (a) Time. Impact refers to loss over a period of time. The longer the period needed to recover from a disruption, the greater the significance of its impact and its lasting effect resulting from its state of inoperability.

A central distribution hub has been lost within your organization. Describe the impact from a tactical, operational, and strategic level.

- (Answer) At the tactical level, the impact is dependent on the severity of the loss (criticality and time) in terms of whether or not the installation will be able to recover and reach its new anticipated level of operations. At the operational level, the central distribution hub, being described this way, would have a significant impact at the operational (regional) level because of the loss of capacity associated with being able to route items correctly (mission). Finally, depending on the volume of business disrupted and the costs associated with the impact, at the strategic level the impact can be measured in the overall loss of performance of the company and the change in volumes of services demanded in the future.

Notes

1. While this is paraphrased, this kind of definition can be easily found on the U.S. Department of Public Safety's homepage for CIP at http://publicsafety. gc.ca/prg/em/cip-en.asp. The safety, security, and economic well-being is actually paraphrased from "Government Security Poicy," Tresury Board of Canada Secretariat, 2002, http://www.tbs-sct.gc.ca/pubs_pol/gospubs/TBM_12A/gsp-psg1_e.asp#pol.
2. Federal Emergency Management Agency, a suborganization underneath the U.S. Department of Homeland Security; http://www.fema.gov.
3. Public Safety and Emergency Preparedness Canada; http://www.publicsafety. gc.ca/index-en.asp. This department has now been renamed as Public Safety.
4. The SPS assists in the movement and coordination of movement of containers within, into, and throughout the shipping process (truck, rail, and maritime). This is (essentially) an automated coordination and inventory control system, similar to the Transportation Shipping Harmonization and Integration Planning System (TSHIPS) by the U.S. Department of Transportation; Nancy L. Nihan, Karl Westby, and Robert Jaeger, "TSHIPS: Transportation Shipping Harmonization and Integration Planning System" (Final Technical Report, TNW2001-03, Vol. 1), March 2001, http://ntl.bts.gov/lib/11000/11100/11178/TNW2001-03vol1.pdf.

Mitigation and Cost Benefit

7

Objective: At the end of this chapter, the reader will be able to define and describe the following:

- The different kinds of risk and what type of risk best fits an organization
- What costs and impact would affect an organization if the environment were disrupted or destroyed (as outlined by the risk determined)
- The difference among the tactical, operational, strategic, and system perspectives in terms of mitigation layers in the transportation system
- How each of these contributes to assuring the services provided by the transportation system

7.1 Introduction

Having identified risks to the organization, steps must be taken to bring risk back to acceptable levels. The organization's management will be able to determine whether the level of risk is acceptable based on abilities of maintaining specified levels of performance. Some regulatory requirements might require organizations to assure that certain goods and services are available. The organization will have to take the impact of requirements into consideration.

7.2 First Step to Mitigating Risk—Strategy

The first challenge associated with dealing with the risks to critical infrastructure involves determining what strategies to take. These will vary depending on the costs, capabilities, and tolerances (with respect to performance) within the organization. For the critical infrastructure protection (CIP) security professional, the challenge is to propose strategies and procedures that flow gracefully among the tactical, operational, and strategic levels.

7.3 Key Considerations

The CIP security professional must be able to make recommendations to management, as would any other advisor. This individual must assess and take into consideration management's various tolerances toward risk, the availability of resources to cover costs, and the resiliency of the organization with respect to change. For the seasoned professional dedicated to protecting critical infrastructure, this may pose a personal challenge as the recommendations put forward are balanced against the needs of the client's organization.

7.3.1 Management Tolerances toward Risk

Certain parts of any organization will operate more smoothly than others. The CIP security professional must act as an advisor by identifying areas in which management feels that the organization is particularly vulnerable or stable. Understanding the management and organizational culture will be critical to determining the success of the CIP program, and the CIP security professional will have to understand the various cultural balances at work within the organization to be successful.

7.3.2 Costs

Cost is usually a driving, if not the most important, factor when looking at a mitigation strategy. The ultimate goal involves appropriately mitigating risk, but this cannot come at the expense of the organization protected. When costs associated with developing, implementing, and maintaining internal programs and capabilities are too high; organizations may choose to have an external organization responsible for dealing with that particular risk. This is when mitigation strategies might cost more than potential impacts and management may choose to accept these risks, particularly when mitigation strategies are not enshrined within legislative or regulatory requirements.

Similarly, the CIP security professional needs to understand that financial risk will also factor heavily in the organization's decision-making processes. When costs cannot be integrated or shared and cannot be linked back into supporting mission lines, the costs will likely be seen as waste should financial pressures increase. This leads to less stability.

7.3.3 Resistance to Change

One aspect that may be overlooked involves an organization's ability to change. Collective agreements can bind organizations and workers in such a way that negotiations must be reopened to implement any changes. Similarly, standard operating procedures (SOPs), if entrenched, may also lead to resistance of change. The simple entrenchment of certain practices or desire to proceed along certain courses of action within the culture of an organization will lead to a level of inertia. The CIP security professional should put forward mitigation strategies that make the best use of existing procedures and capabilities within the organization, seeking to implement an evolutionary rather than revolutionary program.

7.4 Selecting a Mitigation Strategy

Depending on the nature and severity of risk and resources available, management might choose its mitigation strategy from one of the four following options:

1. Avoiding risk: This involves the adjusting of the mission and its supporting business lines so that risk exits the system. Although this may be possible in relatively diverse organizations or at levels with more limited scope, it is generally not possible within the context of the transportation system. The transportation system, at the macro level, has a narrowly defined mission in support of a number of key infrastructure sectors, essentially a situation in which, at the strategic level, the avoidance of risk is impossible.

2. Addressing risk: Sometimes referred to as a mitigating risk, this involves allocating internal resources in such a way that the organization works to lessen the risk. This may involve as little as the adjustment of specific procedures or the establishment of a CIP security program within the organization. Management remains accountable and responsible for the application of those procedures with very little legal recourse should something fail or due diligence be lacking.

3. Transferring risk: This involves allocating internal resources (generally funds) for payment or transfer to another organization that will take steps to mitigate the risk. When it is not cost effective for the organization to establish and maintain a program's capacities, this option may be preferable. In this context, management remains accountable for procedures, but responsibility can be delegated to the organization. The distribution of risk is also considered a method for transferring risk. Under the distribution model, a full community or list of partner organizations becomes responsible for reducing the impact and probability of an event. On a positive

note, this approach builds communities of trust between various organizations. It may also be used inappropriately to avoid accountability.

4. Accepting risk: This generally involves management's determining that it will accept any impacts associated with risk as opposed to accepting the impacts associated with mitigating the risk. This is particularly prevalent where the risk is associated with particularly low impacts. When accepting risk, management affirms its accountability and responsibility regarding any losses associated with that risk.

5. Ignoring risk: This involves management's decision that it will not accept the risk being considered, instead declaring it to be, for example, farfetched or unforeseeable. Ignoring risk means that management remains accountable and responsible not for any losses associated with the risk but may also face challenges regarding the initial decision to ignore the risk.

7.4.1 ABC Transport Example

ABC Transport's distribution warehouse faces a grave risk with the loss of the shipping system due to cyber-related attacks. The CIP security professional hired to provide insight regarding possible losses and impacts associated with this pattern of attack has made a presentation indicating that the loss would close the warehouse facility for a period of 48 hours and would have a significant impact on the remaining facilities downstream from the main facility. He also notes that the risk cannot be avoided because the removal of the system in question would be catastrophic in terms of its impact on the operations in question.

One of the board members from the board of directors (Director A) believes that the company should further centralize controls on the planning system and should hire its own information technology (IT) and security staff such that the organization has resources available at all times to deal with any potential disruptions. This is largely due to the fact that any potential down time would likely cost more than the salaries, operating, and maintenance costs associated with having the people on hand and able to deal with the disruption. In this case, Director A has chosen to address the risk.

Another board member, Director B, disagrees with Director A, noting that the hired IT personnel would have little or nothing to do during periods when a disruption is not taking place. He argues that it is better to establish a contract with a local provider who will guarantee, with insurance coverage, that any disruption will last no more than one hour. Director B has chosen to transfer the risk to another organization.

Director C notes that the network, though connected to the Internet, has many firewalls and other measures and countermeasures already in place and has not appeared to have any disruptions over the past several years. He also notes that, with the awareness of the potential risks, the company would be better served by not allowing additional changes in the system that could open any new vulnerability for attackers to exploit. Director C has chosen to accept the risk.

In addition, the final board member, Director D, remarks that the board has been entirely too focused on network issues and indicates that the board is perhaps a little too paranoid. Director D is choosing to ignore the risk.

7.5 Tactical-Level Considerations

As discussed earlier, the tactical level of an organization has a different perspective from the operational or strategic level. The focus at the tactical level is on the maintenance of performance above specified thresholds. As a result, the prevention, detection, response, and recovery aspects of the security plan will focus on the ability to prevent injury or minimize it. On the other end of the spectrum, the business continuity plan along with an emergency preparedness plan will focus first on preventing conditions but then on minimizing impacts and recovering from them as quickly as possible.

When these considerations are harmonized, facility or tactical-level managers have to answer to the following questions:

1. Is the infrastructure at this location reasonably protected against a determined attack?
2. Does the capability to detect an attack (or other suspicious behavior) exist consistently across the organization and with respect to the work performed?
3. Are people comfortable with how they need to respond if they feel they have detected suspicious behavior or an attack?
4. Does the organization have the capability to report the detection of suspicious behavior or an attack to the appropriate parts of the organization?
5. Does the organization have plans in place to deal with foreseeable or reasonably foreseeable attacks, and does it have in place a mechanism to call together decision makers to deal with unforeseen situations?
6. Does the organization have plans in place to resume operations at an acceptable level as quickly as possible and mechanisms in place to report progress toward reestablishing that level to the rest of the organization?

The tactical level, therefore, looks to understand both what is present and what is interacting within the local or tactical environment. Its focus is on the ability to detect unacceptably high levels of risk and to communicate those to its organization for escalation into the operational and strategic levels.

7.5.1 ABC Transport Example

ABC Transport has decided that the shipping and planning system constitutes infrastructure critical to its operations. As a result, management has asked the facility manager to complete a questionnaire identifying measures that are in place to protect the infrastructure and any potential vulnerability to the system within his office. A security consultant, hired at the operational level, might review the questionnaire and then assist in developing physical, procedural, and psychological measures for the facility. The facility manager will be responsible for confirming that these measures are applied consistently at all times.

7.6 Operational-Level Considerations

At the operational level, the concern focuses on the ability to maintain a level of service within a particular region. Questions posed at the tactical level now shift to encompass a localized, regionalized system. These questions would center on answering the following questions:

1. Is the capacity within the region reasonably resilient, or do single points of failure exist within the system?
2. Does the organization within the region have adequate understanding of its interdependencies and interrelationships that support their ability to deliver services?
3. Does the organization have an adequate understanding of what could potentially signal a disruption in those interdependencies across its breadth and at all levels?
4. Does the organization maintain a level of resiliency within the system (e.g., backup supplies, alternate supply arrangements) so that it can respond to a loss of those inputs?
5. Does the organization maintain the ability to assess and communicate disruptions to these inputs to the strategic level, horizontally to other regions, and back into its own system?
6. Does the organization maintain the ability to monitor its own systems to determine whether these reports are actually being forwarded to the appropriate parts of the organization?
7. Does the organization have plans no only to reestablish its ability to perform but also to bring together the necessary experts and decision makers needed to assess potential impacts across the system?
8. Does the organization have plans in place to prioritize, redirect, delay, or otherwise modify its operations in such a way as to continue to meet demands and minimize impacts at the tactical level while also meeting strategic goals?

At the operational level, management has two goals. The first one involves having systems in place maintaining performance levels of the organization through the integration of resiliency, redundancy, and robustness. This line of effort includes being able to predict, identify, respond to, communicate about, and recover from various levels of disruption (both planned and unplanned) within the system. The second goal involves a level of oversight regarding the performance of the tactical level, certifying that it is meeting the organization's requirement to prevent, detect, respond to, and recover from various kinds of security issues or prevent, detect, mitigate, notify regarding, respond to, and recover from various business continuity and emergency preparedness events.

7.6.1 ABC Transport Example

ABC Transport's Northeast office, which controls one distribution hub and five local offices, receives a monthly update regarding potential areas of concern from the head office. Upon receipt of the update, the regional office assesses these concerns and holds a teleconference call to pass on concerns to management at local facilities. Part of the teleconference call asks whether anything observed at the tactical level could fit within the report. Following an answer, a brief five-minute discussion focuses on what kind of material is needed to disseminate information to workers within the area and to generate material for the tactical level. This includes feedback and what to report. When the operational-level management visits the tactical level, they ask questions regarding informational awareness materials.

ABC Transport also requires all local (tactical) level installations to report where they might rely on a single-service provider, identifying reasons for that arrangement. Depending on the response, operational levels may require tactical levels to (1) identify and enter into an arrangement with a second supplier; (2) include contractual measures ensuring that necessary levels of performance are maintained; or, (3) build additional processes that do not rely on any specific inputs. This information is built into contingency plans so that should a disruption occur ABC Transport can shift its operations, thus minimizing any impacts associated with any given disruption.

7.7 Strategic-Level Considerations

At the strategic level, the questions alter even further from tactical and operational levels. The strategic level has to be able not only to look at the emergence of business threats (competition) and how to respond to those threats but also to monitor strategic environments within which the organization is operating to maintain that level of competitiveness. For the strategic level, the following questions become particularly pertinent:

1. Does the organization rely on one single service or service provider, or is there adequate variety that the organization's business lines would not be at grave risk should any one or a limited number be disrupted?
2. Does the organization maintain the ability to identify interdependencies with its suppliers and clients, and are those interdependencies protected in terms of resiliency, redundancy, and robustness?
3. Has the organization identified possible alternative means of meeting obligations (even if the alternatives involve additional cost for finite periods of time)?
4. Does the organization maintain communication with organizations that would be able to assist in detecting potential sources of disruption in operations or with those that are linked through the interdependencies (e.g., investigative and intelligence bodies, various levels of police organizations)?
5. Can the organization communicate requirements for information throughout the operational and tactical levels, and does it have the capacity to receive and collate responses back from those requests?
6. Does the organization provide adequate support for operational and tactical levels within the organization with respect to their requirements to provide the availability of services?
7. Does the organization have plans or the capability to balance performance within the system (e.g., moving surplus personnel or materiel between various operational levels) during times of disruption to maintain its own level of performance?
8. Does the organization have the capability to detect and recognize potential disruptions, evolving disruptions, and cascading impacts? The goal here is to accomplish this as quickly as possible and before they spread throughout the system.
9. Does the organization have the ability to collate information from across various operational organizations to develop lessons learned based on that feedback? Is it communicated only after proper assessment?

At the strategic level, the focus is on business lines and their abilities to survive. If a tactical-level facility or conduit is failing to perform as required to support the organization, it may be adjusted by the operational or strategic level. However, a facility that is causing harm to the overall organization is generally cut away from the organization.

At the strategic level, the challenge is being able to project potential needs and issues so that resources are available for operational and tactical levels. The main effort for the strategic-level management is ensuring the organization can gather adequate and appropriate lines of trusted information. Management sends this information into the organization to various levels and to persons with particular kinds of expertise so that the organization can be prepared for potential events. At the same time, feedback returned from tactical and operational levels is fed to the strategic level so that a list of options is available.

7.8 System-Level Considerations

While the system appears to be relatively homogeneous, it is actually a system of entities in competition with each other. As a result, cooperation between various organizations is not a natural inclination. The system level (generally the regulatory level) has to be aware of this level of competition by understanding various market pressures.

The question at the system level is whether to restrict or relax the constraints on any activities. This requires an understanding both of how risks associated with these kinds of activities might also shift as those constraints were increased or removed and of how the business activities within the system would be impacted.

ABC Transport has concerns that one of its new clients may make ABC's infrastructure more vulnerable to disruption due to its client's apparent unpopularity. As a result, ABC has contacted, through its regulatory body, the appropriate investigative agencies to determine whether there are any known threats against the client.

With the information received, ABC Transport hires a consultant to present a brief awareness presentation concerning kinds of activities that could indicate if a facility or organization was targeted. Given the threat and risk assessment performed earlier, ABC Transport's management realizes that most of the vulnerabilities to its system are due to its reliance on the centralized distribution nodes and its ties to the shipping and planning system.

The strategic management directs its operational level (regional offices) to identify a backup site that could serve as a temporary distribution hub for a period of up to one month in case of a disruption or operational outage. Once the location has been identified, it needs to be secured and information sent to the strategic level's shipping and planning system administrator who then creates the backup sites within the system. Additionally, regional managers are informed that they are to keep these backups locations a closely guarded secret until 24 hours before they are to go active.

At the operational level, the regional management then begins to identify potential backup sites that could serve as the distribution hub. Initial contracts are set up, with some regional managers putting forward queries to the strategic level as to whether those sites can be exploited for other purposes. At the same time, regional managers contact local law enforcement regarding the situation, requesting any information on potential disruptions. Awareness material is distributed to the various local offices identifying what constitutes suspicious behavior, to whom it should be reported, what details to include, and a suggestion from the local police on some basic steps that can be taken to help build cases (essentially, what can or cannot be said, which is verified by the legal department). These are passed on to the local office.

At the local offices, managers brief their employees (meeting a requirement also under the various workplace safety requirements), review security measures, and conclude that the infrastructure afforded protection. Any reports of suspicious activity are passed on to the operational level, which collates them and forwards them to the strategic level.

7.9 Cost Considerations

Management, in addition to managing risk from a CIP perspective, must also manage a number of other risks, particularly financial risks. Regardless of the protection afforded to an infrastructure, in a competitive environment organizations that fail to make money or generate wealth for their shareholders will fall out of competition and gradually into insolvency. For the CIP security professional, this means working closely with the operations staff and financial staff to work toward solutions that can be linked to as many potential benefits as possible and across as many organizations as possible.

From a planning perspective in which groups must compete for scarce resources, this only makes sense—why compete for resources when, through alliances, both groups can share at least some success? For example, linking access control with both security and measures required under safety regimes regarding due diligence and granting of access can have an impact on the overall costs while maintaining security, safety, and legal risks to the organization. The same principle applies in this context—the key is to maximize the depth (level) and breadth (community) that can participate or benefit from the implementation of the physical, procedural, or psychological measures.

7.10 Benefit Considerations

For the CIP security professional, the first step beyond the review of the threat and risk assessment (TRA) is to meet with other risk-based organizations to determine if any of the risks identified are shared risks that may be alleviated through partnerships. At the same time, meetings with the operations and financial organizations will assist in determining limits that cannot be crossed in terms of the resources available to implement the mitigation strategy. Remember that mitigation is a decision made by management and should (ideally) be well supported by management. There should be some avenue of support and knowledge regarding limits of resources available already apparent in the system.

7.11 Aligning Procedures with Performance

The next step is to determine specific steps associated that will put the mitigation strategy into a practical application. As with any concept associated with total quality management systems (TQMSs), there is a definite value to communicating with all persons involved in a process as it moves from start to finish. In CIP, this focuses on the ability to detect discrepancies between formal and informal processes (documentation versus "on the ground") as well as other aspects that might cause confusion should an incident arise.

Each measure should be tied to how it can support performance levels of the organization from conception to testing and then to normalization. This might require some creativity on the part of the CIP security professional and will very likely require discussions with individuals involved with operations when trying to determine specific shortcomings that can be addressed across an entire system. An example might apply toward access control and supply chain transparency, which may be a good link. Similarly, fire detection and monitoring of controlled areas may also provide some benefit. The key is for the CIP security professional to be in a communication dialogue with those individuals that work in the area to determine what is needed to support them with their work. This may also assist them with gaining the necessary buy-in.

7.12 Setting Strong Procedures

Many regulations are written so that an organization simply has to deter or prevent something from happening. While this is important in terms of understanding where the main emphasis should be, it is generally inadequate from an overall perspective. The key here is to prevent loss, to detect conditions that lead to loss, to notify the appropriate part of the organization to trigger a response, to initiate that response, and then to take steps to get back to acceptable and normal operations. These work in a cycle so that the time it takes to detect and respond to an event is less than the time it takes for the injury to take place.

7.12.1 Prevention

Preventive measures are intended to deter or delay a potential attacker so that the attack is either not worth the effort (in the mind of the attacker) or does not appear to be achievable. At the tactical level, these measures may include fences, gates, barriers, and firewalls.

At the tactical level, preventive measures may slow the throughput of certain processes by adding steps or requiring pauses where checks are completed. The CIP security professional may wish to consider integrating the various steps into existing processes so that minimal additions or changes are made. When lags occur while checks are completed, the CIP security professional may wish to trigger the start of the check earlier so that the step involves having the right information at the right place at the right time.

At the operational level, preventive measures also include planning activities that involve maintaining the level of performance levels between entities at the tactical level. This may incorporate being able to identify any slack within the system that could be exploited during periods of disruption or the identification of suppliers of goods and services that could be called on from the regular supplier if the inputs are disrupted. It may also include recommendations regarding the level of goods and services to maintain in reserve, even if at a cost, based on the time it takes to reestablish the normal supplies.

At the strategic level, preventive measures involve the setting of plans and allocation of resources in such a way that the organization is well positioned to respond during a crisis. These generally include the positioning of the organization so that it is more aware of its own environment and capabilities to deal with that environment.

For example, ABC Transport faces a potential loss of availability of a facility due to a potential disruption. To deal with this situation, ABC Transport has identified persons who possess expertise in its core processes, business continuity planning, event planning, crowd control, and other activities that could be of value. The plan at the strategic level is to shift the roles of those persons into planning and advisor roles should an event occur anywhere in the organization. To ensure a smooth transition, administrative procedures are put in place to have the strategic level pay costs associated with that individual's services and his or her replacement at the tactical level. The repository of business continuity plans in a databank will be expanded so that the full training (both internally and externally) packages for each individual are clearly identified. Thus, the organization knows what internal capability it has to deal with any eventuality and where a person can be contacted on short notice.

7.12.2 Detection

Detection involves the ability to identify and categorize activities, situations, or conditions as being unauthorized or suspicious in nature. There are several aspects to this capability. First, the organization must have a clear sense as to what constitutes authorized behavior and activities, both in a formal (management authorization) as well as informal sense. Second, individuals must have a sense of what information to identify and what boundaries should not be crossed when attempting to gather that information. Third, there must be directions for notification, what, and to whom. Finally, the notification should provide a clear signal or information package so that responders are as aware as they can be regarding the notification.

At the tactical level, detection may involve a range of several abilities. Employee awareness and similar kinds of capabilities provide a significant capacity to the organization. Similarly, technology can provide alarms and signals of a number of varieties that will signal that a change in the environment has occurred. The key at the tactical level is to ensure that there is as much participation as possible within the organization so that any chances of noticing the identified conditions or situations are as high as possible.

In some cases, the rule of "it's suspicious until it can be shown that it's not" may be a good one to follow. This certainly applies to suspicious packages and other items that can cause harm if mishandled. These situations should be identified through employee awareness training programs as well as through consultation with appropriate expertise, often available through local or national law enforcement organizations.

ABC Transport may choose to confirm that its perimeter controls at the distribution nodes are well maintained and to provide a level of control such that plans are in place for guards or personnel at the gates to respond to the appearance of protesters. At the same time, lines of communication are tested, backup sites identified, and other methods of preparations made so that, for the potential protesters, accomplishing their aim is as difficult as possible.

At the operational level, detection not only involves the detection of behavior at the tactical level but also identifies indicators that performance is being affected. This information should be collated in such a way that groups of information with similar characteristics can be identified and communicated laterally across the operational level and to the strategic level of the organization.

ABC Transport may have established a number to call to report suspicious behavior or the presence of the protesters. Employees could be presented a card to be carried with their ID access card that describes what to do, what not to do, what to report, and who to report it to—essentially a security hotline.

At the strategic level, detection involves identifying when events begin to affect the overall organization. This might involve a specific region dropping below a specific threshold or impacts spreading beyond the scope of one region demanding the coordination of the system in general. Depending on the culture of the organization, the management identifies specific conditions that require the strategic level's attention.

When ABC Transport's regional office identifies that the key infrastructure has been disrupted to the point where the impact is spreading to other distribution nodes, notification can be sent from any office and is verified by the strategic management's office.

7.12.3 Response

Notification must lead to an appropriate response, which has three phases. The first phase involves the identification of the scope of the incident and activities that focus on preventing its further spread or injury to the organization. The second phase entails moving beyond the halting of the incident and working toward beginning the recovery process through the stabilization of the environment. Once the environment stabilizes, the third phase will lead to the beginning of the recovery process.

At the tactical level, response focuses on the preservation of infrastructure in order of importance. This will first involve the preservation of life, both within the organization and in the surrounding environment. Once this requirement has been met, the CIP security manager must take steps to attempt to preserve the capacity provided by the organization. This may involve the following:

- Shifting a process away from the affected areas, accepting a reduced loss until recovery can take place
- Shifting the source of inputs involved in a process to reduce the impact of an event
- Generating alternatives so that capacity is maintained through other means

ABC Transport's CIP security manager is considering a response associated with the disruption of the connection to the shipping and planning system. Upon notification of a disruption of the connection to the database, the detailed procedures are put into force. Should the disruption be associated with a loss of the physical connection between the local and regional systems, the local office will attempt to dial into the system at the regional office. Concurrently, the regional office will be informed of the lost capacity.

At the operational level, response focuses on making decisions that maintain the level of performance within the areas. In addition to preservation of infrastructures important to the operational level's activities, this will include adjustments of routes and nodes where possible such that any impact is minimized. To accomplish this, the operational level's knowledge of the location of surplus capacity and capacity shortfalls will be critical.

ABC Transport's regional CIP security manager has identified a disruption to the shipping and planning system's normal connections to the regional office. The strategic level's CIP security manager has been informed of the potential disruption. To maintain the level of expected performance within the region, a connection is established that allows for some communication between the regional and distributed system. Simultaneously, shipments continue to be prioritized based on dates required, value of client, and indicated urgency so that those shipments are completed.

At the strategic level, response focuses on making decisions maintaining levels of performance within the organization while adjusting processes and activities in such a way as to relieve pressure on affected areas. This generally involves balancing and redistributing pressure on the system as to complete the maximum number of transactions in terms of value.

ABC Transport's strategic-level management has received notification that a regional office's connection to the shipping and planning system has been severed and that the regional office is now using its dial-up connection. In discussion with the neighboring regional offices, it is determined that approximately 25% of the shipments bound for that region can be redirected to those neighboring regions, relieving some of the pressure on the affected office. Of that 25%, the company is willing to incur the additional costs associated with the less-than-efficient process for approximately half those packages. The rest will remain positioned at the nearer distribution node.

The purpose of response is to halt immediate injury, to prevent further injury (including containment), and then to set the stage so that the recovery phase can take place.

7.12.4 Recovery

This generally involves a return to normal operations and can take significant time depending on the nature of the event. Having halted the spread of the damage, the next step is to begin to work toward (1) a tolerable level of performance and then (2) an acceptable level of performance.

The first step in this process involves verifying that the response phase has been concluded and that a reasonably stable operating environment has been established. Although many systems might include similar activities as part of the response phase, the recovery phase will need to include some level of awareness building for personnel to make sure that all risks and changes to the environment have been identified.

At the tactical level, this generally involves the conclusion of some kind of verification that any threat has been removed from the system. Once that has been established, personnel and activities are slowly phased back into the process and operations are either resumed or increased gradually. At the same time, the organization should be conducting analyses to determine the root causes of any failures and investigations regarding how vulnerabilities were exploited so that lessons are learned and communicated to the operational level.

ABC Transport has reestablished a connection with the shipping and planning system. Having reestablished that connection, reviews of transaction logs and other access control logs are being reviewed to ensure that no unauthorized modifications were made or suspicious activity is present. Following that review, the shipping and planning cell's systems are connected in coordination with the central system to certify that the reestablishment of services is controlled and reasonably graceful.

At the operational level, the connection's reestablishment means that a significant benchmark in the response phase has been met. The recovery plan has a series of goals, checks, and balances that are to be met to assure that the system is not put under undue strain.

At the same time, regional management communicates the updated status to the strategic level, informing that level of an estimated time of return and any plans to reestablish normal operations. Operational management is looking into root causes of the disruption and attempting to determine why the system failed and what can be done to prevent future recurrences of the event.

ABC Transport's regional office has reestablished its links with the strategic and local offices. To return to the planned schedule, additional trucks and shifts are added to the system so that the deliveries are made on time. Connections are verified during set periods of system down time.

While recovery occurs, the operational-level's management lets the strategic-level management know the estimated time for returning to an acceptable level of operations. Operations also provide a report regarding the causes of the failure and measures to prevent future occurrences.

7.13 Linking Business Activities

Within the business community, the system-level mitigation strategies and measures have three major efforts. The first effort deals with aspects of prevention or taking those steps necessary to reduce the risks associated with being unprepared for an event. This refers to the preventive and preparatory measures that are generally associated with business continuity planning, contingency planning, and emergency preparedness. Second, measures associated with detection within the security realm refer to the concepts of detection and notification within the business continuity and emergency preparedness realm. Response and recovery tend to be interrelated between the two systems.

For the CIP security professional, this provides an opportunity for a third effort: harmonization within the three systems. Combining these systems with the safety management, environmental response, and pollution control systems can further extend the reach of the program while also allowing each individual effort to benefit from greater levels of efficiency. The challenge is in developing the necessary tables of concordance that would likely be required should the organization undergo a comprehensive audit in any single program.

7.14 Robustness, Resiliency, and Redundancy

These three terms generally indicate the ability of any given system to weather some kind of disruption. When looking at the overall transportation system, the ability to instill robustness, resiliency, and redundancy into that system is a highly desirable goal.

7.14.1 Robustness

This indicates the ability of a specific entity to withstand or absorb damage. At the tactical level, this generally involves protecting infrastructure in such a way that it takes a considerable attack to damage it. At the operational and strategic level, robust systems generally have significant protective measures but also include significant levels of resiliency and redundancy that allow the systems to survive considerable damage.

7.14.2 Resiliency

This deals with the ability of something to reestablish itself. At the tactical level, this involves the ability of infrastructure to be repaired or replaced so that the down time associated with that infrastructure is minimized. At the operational and strategic level, this deals with the ability of the system to shift and then to reestablish its normal routes and conduits.

7.14.3 Redundancy

This indicates conditions where duplications of services, or equivalencies, exist within the system so that where there is a loss in the system the requirement simply shifts onto another person, asset, facility, information, or activity in the system. At the tactical level, spares or replacement capacity exist at the location or are able to be called to the location. At the operational and strategic levels, this involves the ability to use surplus capacity in the system to accomplish the same goals being met before the infrastructure was lost so that performance is not diminished system-wide.

7.14.4 Cascading Impacts

These will have influences across all levels of the system. At the tactical level, changes in infrastructure that create losses or increases in capacity will influence the capacity available at the operational level. At certain points dependent on the network, these increases or losses will influence the impact across the system. Depending on the extent of the disruption at the tactical level (in terms of location, volume, and extent), the stability of the operational level will be influenced. As this influence permeates the system, it will eventually affect the strategic level. In essence, disruptions within the network will likely cause a level of fragmentation and finally dissolution of the network that can only be addressed through the reestablishment of capacity and conduits. For the CIP security professional, understanding this ebb and flow of risk as it pertains to the probability of disruptive impact (leading first to disruption, then to fragmentation, and finally to dissolution) is key.[1]

7.15 Setting Goals and Benchmarks

When looking at the goal-setting and benchmarking exercise that will determine the level of potential risks and impacts existing in the organization, it is important to use a system (in the context of a coordinated effort) that reflects the mission of the transportation system. Throughput provides the tool that allows for these measurements.

The first method involves tracking the loss of capacity that describes an impact in terms of survivability and performance. This applies to the tactical, operational, and strategic levels of the organization. It may be applied to the transportation system to sustain the economy of the nation, etc., when one looks at the level of economic performance required. It is important to note that this system can be used in terms of loss (unable to perform) and of the impact of security measures (additional time required leading to fewer transactions per unit of time). The second context involves the reduction of risk that may be determined through the calculation of reasonable outcomes with respect to loss. As discussed earlier, risk is a factor of probability and impact, and the reduction of either probability or impact reduces the level of risk in the system.

7.16 Generating the Manual

Over time, as the organization and systems mature a manual of lessons learned will be compiled. Where success is realized, the root-cause analysis identifies the peculiarities that allowed that success to be achieved. In each case, the key is to reduce, through root-cause analysis or a similar mechanism, the steps taken that directly linked a mitigation approach to a vulnerability to a risk and then to the environment in which it is operating. For example, the success associated with one office preventing theft by leaving the lights on would be reduced to the recommendation that an area be illuminated, eliminating the ability to operate from concealment and therefore reducing the risks associated with theft. As the organization and system matures and as successes and failures are tracked, the manual grows in such a way that it evolves into a series of tools available to those in the system.

7.17 Questions

Based on the various levels of perceived risk, which risk might have the most significant impact to any given organization? State your reasons for choosing your answer.
- This depends on the management culture. Risk is generally defined as probability and impact, but perceived risk deals with a number of influences, many of which can be unique to the individual.

Do legacy infrastructure environments offer more or less toward resistance and resiliency to change? Based on your answer, provide an example of how it would affect your organization, and what factors would play a part in such a movement. What factors would need to be present to eliminate such factors?
- (Considerations) Legacy systems may provide a layer of resistance to change in terms of their entrenchment but may also offer opportunities for sudden incorporation of resilient systems as management realizes that the infrastructure requires upgrading. Similarly, scalable systems and modular systems may also offer opportunities in this respect.

Define how throughput and risk correlate with each other. Under which circumstances would they be related to each another, and conversely, when would they not be related?

Describe how a decision to close a local facility in your organization would affect your overall company. Now consider the closure of a more significant local office. Write down the impacts at the neighborhood, city, regional, and national levels beside each other. Note factors that trigger the escalation of risk in the system.

Notes

1. This concept is generally thought of in terms of information networks. The transportation system, however, can be looked at in terms of a complex and adaptive network (as discussed earlier) with economic factors providing the keys to motivation. The concepts associated with disruption and adaptation are put forward in such works as John Holland, *Hidden Order: How Adaptation Builds Complexity* (Reading, MA: Addison-Wesley, 1996); C.S. Holling, "Understanding the Complexity of Economic, Ecological and Social Systems," *Ecosystems,* Vol. 4, no. 5 (August 2001), 390–405.

Certification, Accreditation, Registration, and Licensing

8

Objective: At the end of this chapter, the reader will be able to define and describe the following:

- The differences among certification, accreditation, and registration
- How a continuity of operations plan would fit within an organization
- How each of these contributes to assuring the services provided by the transportation system

8.1 Introduction

Having identified risks and steps taken to mitigate those risks, the question remains as to whether steps have been taken to satisfy various levels of authority within the system. For those who have an information management and information technology (IM/IT) security background, the concept of certifying security measures and accrediting participation in a networked context will be relatively straightforward. For those outside this community, this describes a series of efforts that intend to manage risk in an auditable, accountable, and actionable system. Here the same concept applies to the context of the transportation system.

Certification, accreditation, registration, and licensing provide pivotal roles across and throughout the transportation system. Remember—the transportation system tends to allow infection to spread through itself, much like the human body. Unlike the human body, the transportation system is based on a collection of competing entities. This means that the chance for the system to arrive in a state of harmony is reasonable, but the chance for the system to arrive at a state of cooperation, at the expense of competition, is unlikely.

8.2 Linking to Mitigation

Mitigation involves establishing physical, procedural, logical (technical), or psychological controls to reduce the levels of risk acceptable levels. When private-sector entities are involved, the decision to avoid, address, transfer, accept, or even ignore the identified is generally traceable to the financial and operational values of organizations. The decision to avoid risk through the changing of operations would be represented by entities choosing to leave the system. This would signify a situation in which the participants in the system (who are motivated by the generation of profit) conclude that the environment will not allow that to happen. The decision to accept or ignore risks is limited by the regulator of the system compelling participants within the system to either address or transfer risks that the regulator concludes are beyond social tolerance. These constraints are in addition to the various social and cultural norms that would constrain the entity's operations within a community. This causes friction between the regulator and the industry as the former strives to limit the activities of the latter while the latter strives for the most flexibility attainable to continue generating wealth.

8.3 Certification

Certification, particularly certification based on common criteria, is the judgment by experts that certain controls (i.e., physical, procedural, logical, and psychological) are of suitable strength that they offer an identified level of protection. This concept has been around for some time and is used by audit and standards communities. However, it is only beginning to make some headway into the security and critical infrastructure protection (CIP) domains.

At the system level, the overall community begins the process by identifying frameworks and baseline requirements that must be present across the entire system. In the case of international organizations, such as the International Maritime Organization (IMO) or the International Civil Aviation Organization (ICAO), these requirements are taken back to various nation-states that choose to either become signatory states by signing the document or to remain outside of the community. The signatory states then enshrine requirements within their legislative structures refining them through regulations to define their mandatory requirements—limits of what risks may be accepted or ignored. These baseline requirements are communicated across the community so that they apply within their various administrative and geopolitical boundaries.

Private-sector entities may add additional controls where they feel it is appropriate, as long as those controls do not interfere with the legal and regulatory requirements imposed on them. Senior management within an organization can choose to impose additional controls through their own internal policies, standards, procedures, and guidelines. They can communicate baselines in terms of limits of delegated authority given to each manager. This would require the manager to seek higher levels of authority when making decisions that could expose the organization to higher levels of risk.

Certification results in a statement that all requirements within a system meet established criteria and, in an expert's judgment, are sufficiently strong enough to be considered effective. There are four key elements to this statement:

1. Criteria must be established, stable, and communicated so that they may be evaluated.
2. The person assessing the level of compliance must be an expert who can render appropriate judgments that the controls are appropriate within that context.
3. The full suite of requirements must be both considered and acted upon.
4. The controls put in place must be suitably rigorous to assure first the expert and then the accreditation body that they are suitable given the environment.

The certification of the entity (be it a node or a conduit) provides a way entities can be compared across the entire system.

8.4 Accreditation

Whereas certification indicates that certain measurable standards have been achieved, accreditation provides management with a review mechanism relevant to the specific risk environment. Consider, for example, that one entity is in an area fraught with peril while another operates within a tranquil and supportive environment. Both have achieved certification, indicating that they have met certain goals. Accreditation shows how length of time the certification is valid, any conditions for maintaining it, and under what conditions it can be revoked.

The concept of accreditation provides management with the first layer of risk management from an infrastructure assurance and continuity of operations point of view. Certification establishes that an entity has reached baseline expected across the system, allowing the overseeing body a level of comfort that the system is being protected. Accreditation allows that oversight body to set the level of rigor that will be applied in ensuring that the certification's requirements are being maintained at all times. This is accomplished through the following:

- Adjusting the period for which certification remains in force, essentially requiring the subordinate parts of the organization to be under additional monitoring and oversight
- Adjusting the requirements to be met as a condition of certification to meet specific or unique risks within the system from both the infrastructure assurance and capacity maintenance perspective
- Indicating, through official communications, the requirements and opportunities afforded by the accreditation

8.5 Registration

Registration includes the entity in what is considered a *trusted system*. This system is a collection of individual decisions throughout the accreditation process that gradually completes the understanding of threats, risks, vulnerabilities, and mitigation strategies across the overall system. The consistent and harmonious response to risk issues that characterizes this community allows for a level of comfort within that community and its stakeholders.

Registration serves two purposes within the transportation system. First, it makes certain that basic levels of infrastructure assurance and capacity management have been maintained in the system (from regulatory and corporate perspectives). Second, it involves an understanding of system stability that comes from the ability to monitor the population of the system. In this respect, the registration process of the transportation system is not yet mature, as key information in the accreditation process is considered sensitive from a competitive point of view.

This missing information poses a system-wide vulnerability, particularly since the focus of regulatory-based systems shift in their focus from the accreditation process onto the registration process. Security has become a marketable commodity in which an entity appears to be less secure, there is a clear concern that this can lead to business losses.[1] This can put pressure on the overseeing body to give special consideration or make exceptions, and these may allow for a level of contamination to enter into the system. Thus, the registration process generally involves communication of the lists of registered entities and not by their accreditation; rather, their inclusion on the list assumes that the accreditation has been met.

8.6 Licensing

Licensing within the context of the transportation system involves the formal communication between various levels of oversight and entities stating that a business can legitimately carry out activities as long as it operates within the confines communicated during the accreditation process. This includes a tacit acceptance that violating the conditions of any of these licenses can result in sanctions taken against the noncompliant or violating entity or its management.

Although this concept is understood when the entity is dealing with the regulator or some other government office, it does offer opportunities for the private sector. The key difference is that private-sector entities create a licensing agency for the express purpose of carrying out specific operations, whereas the regulator monitors conditions at the entity to determine if accreditation can be achieved and its registration processed.

8.7 The Trusted Transportation System

Given that there are two goals—the (1) assurance of infrastructure and (2) the maintenance of capacity—the concept of certification, accreditation, registration, and licensing plays a significant role by identifying what can be considered the trusted network. The value of participating in this network is associated with the understanding of how it responds to risk and the comfort associated the knowledge that baselines are consistently applied throughout the network. This does not mean that a particular node or conduit is immune to any threat or risk, only that the system has confidence that all reasonable or appropriate measures have been taken by all participants to protect the capacity and infrastructure within the network.

During the initial phases of the assurance process, we identified the capacity of each entity in the system and used those capacities to determine how critical a person, asset, facility, piece of information, or activity was within a process. Second, we looked at those nodes and conduits and how they interact within the tactical, operational, and strategic levels. Now consider that we have identified what nodes and conduits considered to have adequately mitigated risk through the certification, accreditation, and registration and licensing of the system. Given we have also categorized the vulnerabilities within the system (allowing for the ability to categorize nodes and conduits on those bases), we can also begin to identify possible areas that could fall prey to the same threat agents given a particular kind of attack. At this point, this knowledge forms a picture of the known and trusted transportation system. All entities outside of this trusted network must undergo some kind of verification that they do not pose an unacceptable risk before being able to interact with the trusted network.

A key challenge in maintaining this level of awareness involves establishing and maintaining the ability to communicate information across the network. This will first involve identifying and then managing and maintaining a system involving the categorization of data within all components of the system. This categorization involves the establishment of a common base of definitions. Key terms may not be defined the same since specific domains have been allowed to operate in isolation of each other. Without this common base, the internal communications will always be subject to confusion.

The second involves the mechanics of communication and the interoperability of those systems. Even if the message is clear, if there is no mechanism by which it can be communicated, then effective communications cannot be assured. How that message is identified, routed, and made available across the system is another challenge. In this respect, the categorization of data so that it aligns with the various business lines and missions becomes essential so that the trusted networks can communicate effectively within the trusted system. This is accomplished either directly or through trusted translation processes operating in trusted environments. This requirement is key toward the establishment of a system that will be able to identify challenges, to define the scope of those challenges, to formulate plans and measures, and to monitor those responses to risk to determine whether the system is functioning as intended.

8.8 ABC Transport Example

Strategic management of ABC Transport has determined that they want to establish a reputation for reliable deliveries in support of their mission to deliver services within a key market niche. Communications has determined key messages that would be of significant benefit in support of this effort by demonstrating that the organization can deliver services during periods of increased difficulty (primary message) while preserving those services in the longer term.

ABC Transport needs the following key officials to provide support to the initiative:

- Chief operating officer (COO): Can services be maintained during periods of increased risk or difficulty?
- Chief financial officer (CFO): Can we sustain current levels of operation yet remain viable as an organization?
- Legal counsel: Can we make this claim legitimately, and what constraints should be built into the system to protect us from legal risk?

The COO turned to the head of business continuity planning (HoBCP), a subset of the security organization, to determine how the organization would react to periods of disruption to its key personnel, assets, facilities, information, and activities. The HoBCP indicated that a certified business continuity planner had reviewed all facilities using plans based on sound business continuity planning practices and guidelines. Requirements for testing, drills, and exercises were part of these plans, although such activities were often cut short due to personnel and operational issues.

The COO turned to the head of corporate security detail (HoCS) to determine if the infrastructure could be considered protected. The HoCS indicated that each distribution node had undergone a significant review of its security posture by a certified protection professional and other competent security professionals. Local offices had not completed the process.

Following interviews with organizations responsible for occupational health and safety, environmental response, and pollution control, the COO realized that each facility was reasonably compliant with the corporate policy but that coordinating the overall systems was a challenge. He then held interviews with each of the five regional directors to determine to what extent they could maintain a level of performance within their region. Only two out of the five had plans as to how to deal with the loss of a local office, and the system was relatively weak with respect to a loss of a distribution node.

Each regional office was then charged with identifying whether it was operating at full capacity. By identifying where excess capacity existed within the organization, the COO operating began drawing up plans involving prioritization of shipments and movement of those shipments using the excess capacity as much as possible. Additionally, the chief security officer (CSO) indicated key areas where additional capacity was needed either to generate the necessary flexibility in support of the plan or to reduce instances where local offices were operating beyond their capacity. Plans were drawn up to identify how to achieve this excess capacity on short notice (Figure 8.1).

8.9 Continuity of Operations Planning

By identifying each node and conduit in the system, the organization can begin to plan a number of key activities to ensure the continuity of its operations. As noted earlier, this involves being able to prevent, detect, respond to, and recover from incidents that threaten the capacity of the system. These include the following:

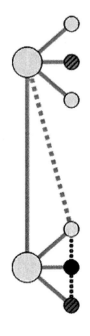

In this region, two local offices have redundancy and resilient ties back to their distribution node. Each grayed office is operating around 95% capacity in relatively stable communities. The certification/accreditation process has ensured that the two offices are operating with appropriate protection and continuity plans in place. The hashed-grayed office operates in a higher risk area, requiring additional preventive measures and continuity options.

The hashed-grayed office is operating in a higher-risk environment at the same capacity. As a result of this, ABC Transport's certification and accreditation process requires that it undergo a higher level of review. Additionally, the Chief Operating Officer has directed the Regional Manager to develop a plan that would prioritize shipments bound for that office to be redirected through the other two offices (using their surplus capacity) so that deliveries could be maintained while the local office moves to its alternate location and reestablishes its level of service.

This region faces and additional challenge as one of its offices is failing to make its performance standards (black facility), bleeding resources out of the regional pool to support it. Similarly, another office is operating near capacity but cannot tolerate a significant disruption (hashed-grayed).

In this instance, the Chief Operating Officer recommends that the Regional Manager realign the Internal boundaries to distribute the performance and capacity more evenly across the system. At the same time, the Regional Manager is advised to draw up plans that would prioritize the shipments to the flanking regions.

At the same time, the Chief Operating Officer has identified that each distribution node is operating at approximately 85%. %. As a result, a conduit is established between the two nodes that would allow for one to stand in for the other during periods of disruption and for a period of time that would allow for the other to establish its back up site. This involved either sending urgent packages directly to the other regions local offices or acting as a holding point outside of the affected area until such a time as the affected area could reestablish its level of services.

Figure 8.1 Safeguarding capacity.

- Identifying and communicating the indicators and warnings that point to events that may lead to a disruption of capacity
- Identifying options available to maintain the conduits into and out of nodes in the system (e.g., police assistance)
- Identifying options available to make sure that priority of services due to the ability to deliver emergency or recovery supplies (e.g., priority dial tone)
- Identifying options to redirect services around disruptions in the movement using lines where excess capacity exists in the system to minimize disruptions to the overall system
- Identifying thresholds when tactical offices will be required to relocate and reestablish acceptable levels of services
- Identifying how to prioritize the shipments received by the system and directing them in such a way so as to minimize pressure on organizations seeking to reestablish services
- Identifying specific controls that could be either strengthened or relaxed to provide the smooth flow of material in the system

Continuity of operations planning involves the ability to identify excesses and shortages of capacity exist and the ability to match resources necessary to address those conditions. Having identified these conditions, the organization must be able to communicate this information in such a way that the organization's resources can be applied appropriately to the situation. Individuals with specialized skills may need to assist in particular circumstances or technology may need to be made available so that groups of experts in diverse ranges of fields can be brought together (e.g., teleconferences, videoconferences, meetings, symposia).

Having established the ability to communicate, the next step in the process involves identifying the level of capacity that resides within the system. This involves being able to identify and qualify the following:

- The potential attainable level of performance and what is needed to reach those levels
- The actual sustainable level of performance
- The thresholds associated with each level
- The threshold associated with the introduction of loss or the loss of viability
- The specific mitigation plans and the vulnerabilities they address
- The specific indicators and warnings associated with potential disruptions and how to detect those both locally and from outside the organization

Incorporating these requirements into the certification and accreditation process begins to allow the organization's management to identify the most stable system allowing them to continue to deliver their services.

The concepts of continuity of operations planning and business continuity planning both refer to the need to maintain capacity within the transportation system. This capacity provides the link between the sectors that would cause the flow of impact between the sectors along interdependent lines.

8.10 Questions

1. Does the certification process refer to individual or organization certification? Whatever your choice, how would that differ from the other type of certification?
2. How are the two types of certifications similar to each other, if at all?
3. In which aspect does continuity of operations planning have a vital role (e.g., accreditation, certification, registration, licensing, and combinations of these)? Why?

Notes

1. These business losses come from two sources. The first are the direct business losses that are involved with the destruction, damage, or loss of property or value. The second involves the removal of the organization from the trusted system. For example, the IMO publishes lists of ports that are considered compliant with its security regime. If a port does not appear on this list, there is often concern that using one of these ports could result in delay due to inspections, another form of waste as the ability to deliver on time may be affected.

Continuity of Operations

9

Objective: At the end of this chapter, the reader will understand the following:

- The concept of continuity of operations (COOP) planning (similar to business continuity planning [BCP]) and how it relates to contingency capacity planning of the transportation system
- Fundamental concepts surrounding the reasons for implementing COOP and BCP contingency plans
- Methods for implementing response and mitigation and recovery of the system if degraded or compromised
- How COOP and BCP contingency planning and supply chain management are an integral part of the system

9.1 Introduction

The transportation system's value, as a provider of a key service across so many other critical infrastructure sectors, lies not in the protection of a single asset but from the system's capacity to deliver services. This relates directly to activities such as BCP and COOP. For the critical infrastructure protection (CIP) security professional, the sole valid approach to transportation security is to look at the entire transportation system and not just at any one particular node or piece of infrastructure.

This closely parallels efforts currently under way with supply chain management security, but with a few notable exceptions. The supply chain deals largely with the movement of goods, whereas this book deals with assurance of capacity to move anything between one or more points. Interdependencies associated with the transportation system are such that many of these work at the level of symbiosis.

151

It should be noted that the terms *business continuity planning* and *continuity of operations planning* are parallel to each other; BCP is used mostly by government, business operations, and private-sector industries in Canada, whereas COOP is used primarily by the U.S. government, although the concept has moved into the private sector. Countries other than the United States often use the term BCP. In a macro sense, the seasoned and broadly trained professional will want to look at BCP, incident response planning (IRP), disaster recovery planning (DRP), business resumption planning (BRP), and security management and how they relate to each other across a spectrum of mutually supporting activities. CIP intends to build the lines of communication between these and other risk-based systems so that they can work in an efficient and harmonized manner.

9.2 What Is COOP?

COOP is the policy of many worldwide governments to have in place a comprehensive and effective program to ensure the continuity of essential organizational and governmental functions under all circumstances. These plans represent a baseline of preparedness for a full range of potential emergencies for all governmental departments, organizations, ministries, and their agencies. The plans represent a viable planning capability that makes certain that the performance of essential functions during any emergency or situation that may disrupt normal operations.[1]

In the wake of the September 2001 terrorist attacks in the United States, subsequent anthrax incidents, and occasional warnings of potential terrorist reprisals in response to U.S. interventions in Afghanistan and Iraq, U.S. policymakers have given renewed attention to continuity of operations issues. COOP planning was originally developed by the U.S federal government as a segment of government contingency planning linked to continuity of government (COG). Taken together, COOP and COG were designed to provide survival of a constitutional form of government and the continuity of essential federal functions.[2] Within the executive branch, COG planning efforts focused on preserving the line of presidential succession by safeguarding officials who would succeed the president of the United States in a predetermined order.[3]

COOP planning refers to an internal effort of an organization, such as a branch of government, department, office, or ministry, to assure that capabilities exist to continue essential operations in response to a comprehensive array of potential operational interruptions. Although much of the renewed impetus for COOP planning focuses on responding to potential attack, operational interruptions that could necessitate the activation of a COOP might also include routine building renovation or maintenance, mechanical failure of heating or other building systems, fire, inclement weather, or other acts of nature.[4] Failure of information technology (IT) and telecommunications installations due to a malfunction or cyber attack are examples of other events that may interrupt an organization's activity.[5]

Within any given organization or government, COOP planning might be viewed as a business operational continuation policy of a basic emergency preparedness plan[6] or as a bridge between planning efforts to maintain continuity of an organization in the event of a significant disruption to its activity or its subordinate organizations.

In the aftermath of an incident, initial efforts typically focus on safeguarding personnel and securing the incident scene. Subsequently, attention focuses on reestablishing critical operations according to a COOP plan. Because the number and types of potential interruptions are unknown, effective COOP planning must provide, in advance of an incident, a variety of means to assure that contingent operations exist.[7] Law, policy, and organizational plans specify arrangements for the contingent operation of the organization's executive branch in the event of a regional or national emergency, catastrophe, or other operational interruptions. These sources identify a number of matters that COOP planners must incorporate into their planning. In practice, the specialized nature of the various entities of the organization's executive branch results in COOP planning that is highly decentralized, with each organization or division developing specific plans[8] appropriate for maintaining its operations in an emergency. The growth and change of mission-critical needs, personnel, and information systems also help to drive COOP planning.

9.3 Aligning COOP, BCP, and Contingency Planning

COOP grew from efforts established during the cold war to preserve the operational continuity of government in the event of a nuclear attack on the United States. BCP made its most significant appearance in Canada as part of Year 2000 preparedness efforts.[9] Contingency plans within the transportation system stemmed first from operational needs, then as a byproduct of several safety regimes, and finally from either regulatory requirements or the threat and risk assessment. To understand the direction within the transportation system, a brief history of COOP is in order, as the term is only beginning to enter into the vernacular within the various industries.

Following the basic security procedures, two programs work to ensure that the organization continues to maintain its capacity when disruptions (incidents) are first detected. The first involves BCP that balances with the IRP organization. The second level of this involves the DRP and the BRP organizations. Contingency planning generally involves an enterprise-wide approach to potentially grave events and would therefore align in parallel with the DRPs normally associated with the IM and IT domains. Although many organizations have only recently started to address these issues, it cannot be discounted that many organizations, public and private, have had comparable, if not parallel, systems in place.

9.4 Background of COOP

COOP planning is simply a good business practice,[10] which is part of the fundamental mission of agencies as responsible and reliable public institutions. For many years, COOP was conducted at an individual agency level, primarily in response to emergencies within the confines of that organization. The content and structure of COOP plans, operational standards, and interagency coordination (if any) were left to the discretion of the agency conducting the planning.

The changing threat environment and recent emergencies, including localized acts of nature, accidents, technological emergencies, and military or terrorist attack-related incidents, have shifted awareness of the need for COOP capabilities that enable agencies to continue their essential functions across a broad spectrum of emergencies. Additionally, the potential for terrorist use of weapons of mass destruction (WMD) has emphasized the need to provide the president of the United States with capabilities that establish that the continuity of essential government functions exists across the federal executive branch.

COOP planning is carried out under the authority of classified security presidential national security directives as well as through publicly available presidential executive orders. The current documents governing contingency planning activities include Presidential Decision Directive (PDD) 67, Enduring Constitutional Government and Continuity of Government Operations, Executive Order (E.O.) 12656, Assignment of Emergency Preparedness Responsibilities, and a series of Federal Emergency Management Agency (FEMA) guidelines, providing guidance for preparing and exercising COOP plans.[11]

To supply a focal point to orchestrate this expanded effort, PDD-67[12] established FEMA[13] as the executive agent for the federal executive branch COOP. Inherent in that role is the responsibility to formulate guidance for agencies to use in developing viable, executable COOP plans, to facilitate interagency coordination as appropriate, and to oversee and assess the status of COOP capability across the federal executive branch.[14] Additionally, each organization is responsible for appointing a senior COOP planning executive as an emergency coordinator to serve as program manager and agency point of contact for coordinating organizational COOP activities.

In response to the presidential directive, some organizations formed task forces made up of representatives familiar with organizational operations and contingency planning. These plans identified essential requirements to support the primary function of an organization, such as emergency communications, establishment of a chain of command, and delegation of authority. The full text of PDD 67 is classified, with no official summary or other information about the directive released to the public.

9.5 Objectives

COOP planning is an effort assuring that capabilities exist to continue essential organizational functions across a wide range of potential emergencies. The objectives of a COOP plan include the following:

- Ensuring the continuous performance of an organization's essential functions and operations during an emergency
- Protecting essential facilities, equipment, records, and other assets
- Reducing or mitigating disruptions to operations
- Reducing loss of life and minimizing damage and losses
- Achieving a timely and orderly recovery from an emergency and the resumption of full service to customers

Many of the COOP programs will continue to evolve as COOP policies, requirements, and procedures change. New national COOP programs incorporate lessons learned from real-world events and training activities, improving the ability of government to deliver essential functions under all conditions.[15]

   BCP objectives closely mirror COOP objectives in that they attempt to maintain the performance of the organization—after, of course, the protection of lives. Within BCP processes, the following direct parallels may be found:

- Identifying, protecting, and monitoring the core business services of an organization
- Taking steps to protect "critical" persons, assets, facilities, information, and activities and harden them within the context of an all-hazards approach
- Identifying steps taken to minimize the impact of events and with the specific aim of reestablishing an acceptable level of performance as quickly as possible
- Ensuring that lessons learned from one event are communicated appropriately across the organization to further reduce the risks associated with similar kinds of losses

Contingency planning, in its application, tends to focus on events that have a much higher consequence or gravity. COOP and BCP focus on an all-hazards approach, which involves adopting a program that is prepared, through robustness and resiliency within the program structures, to identify, adapt to, and finally thrive within its new environment. Contingency planning, on the other hand, tends to focus on narrow events of significant consequence or gravity and is used to deal with the command, control, communications, and information requirements associated with getting the organization through a particularly nasty but finite event.

Contingency planning also has an element associated with public confidence. Whereas COOP and BCP assure various stakeholders, clients, and populations that the organization is robust and resilient, contingency plans tend to demonstrate that the organization is prepared for any type of an event. One could argue that contingency planning is reactive to the environment whereas COOP and BCP are preventive in nature.

9.6 Elements

The specific details of a COOP plan will vary from organization to organization; FEMA's guidance suggests that executive branch COOP planners incorporate several common components within their COOP planning. These components include the ability to maintain any plan at a high level of readiness, capable of implementation both with and without warning of an interruption of routine operations. FEMA suggests that COOP plans should be operational no later than 12 hours after activation and that they provide for sustained agency operations for up to 30 days.

COOP planners should take maximum advantage of existing agency field infrastructures to the extent that such facilities are available. FEMA recommends that organizations develop and maintain their COOP capabilities using a multiyear strategy and program management plan. This plan would outline the process that the organization would follow to designate essential functions and resources, to define both short- and long-term goals and objectives, to forecast budgetary requirements, to anticipate and address issues and potential obstacles, and to establish any planning milestones. A completed COOP plan would likely incorporate several elements such as the following:

- Identification of an organization's essential functions, which must continue under any circumstances
- Stipulation of organizational lines of succession and delegation of authorities
- Provisions for the use of alternate facilities
- Establishment of emergency operating procedures
- Establishment of reliable, interoperable communications
- Provisions for the safekeeping of vital records and databases

- Provisions for logistical support
- Personnel and staffing issues
- Security measures for personnel, records, and alternate facilities
- Development of exercises and training programs to assure the effectiveness of the COOP planning process

BCP, in its organizational sense, follows similar patterns of activities to assure the most efficient recovery. A symbiotic effort between those with specialized planning knowledge and the organizations actually delivering services best suits the planning process. Thus, the BCP organization (often with the support of security and other organizations) begins to shift away from specific directions to being an enabling, coordinating, and information-sharing organization for the programs and operational elements of an organization. BCP is about providing the knowledge that necessary infrastructures are in place for the mission-critical functions of an organization to continue at acceptable levels.

Contingency planning tends to focus on specific requirements (as outlined already). A BCP organization will generally respond to needs derived from the programs and operations elements. Within the contingency planning realm, issues associated with unity of command and timeliness of communications take priority, meaning that the contingency planning organization takes control of the situation and then works to normalize the environment as much as possible.

The sensitive nature of contingency planning and the specialized nature of government agencies are factors in the lack of publicly available detailed agency-by-agency information regarding the extent of COOP planning. In the winter of 2001–2002, President George W. Bush issued several executive orders providing for an order of succession in the executive departments and the Environmental Protection Agency.[16] Some agencies within other departments have established leadership succession contingencies as part of their COOP planning process.[17]

9.7 Operations

A COOP plan includes the deliberate and preplanned movement of selected key principals and supporting staff to a relocation facility. As an example, a sudden emergency, such as a fire or hazardous materials incident, may require the evacuation of a facility with little or no advance notice, but for a short duration. Alternatively, in an emergency so severe that an organization's facility is rendered unusable for a period long enough to significantly impact normal operations, COOP plan implementation may be required. Organizations develop an executive decision process to review of an emergency and to determine the best course of action for response and recovery. This precludes any premature or inappropriate activation of an organizational COOP plan.[18]

One approach to certifying that a logical sequence of events in implementing a COOP plan is time phasing is as follows:

- Phase I: *Activation and Relocation*—time frame: hour 0 to 12 from emergency
- Phase II: *Alternate Facility Operations*—time frame: hour 12 to termination
- Phase III: *Reconstitution*—time frame: termination of emergency activities, followed by a return to normal operations

BCP follows similar kinds of structure but becomes prescriptive only where and when required to do so. Although this makes the BCP process more rigorous in terms of effort, the net benefit for such a price becomes apparent when one looks at the ability to integrate resiliency into the system.

9.8 Issues Implementing COOP

Policy questions and issues sometimes arise as organizational executives might examine the status of COOP planning within the executive branch of organization (or government) and the implications of that planning for overall organizational emergency preparedness. Some of the issues regarding COOP planning are as follows:

- Immediacy and timeliness: As the memory of dramatic interruptions such as the September 11, 2001, attacks and anthrax incidents fade, attention to administrative operations like COOP planning may receive lower priority attention from agency planners. Emergency preparedness observers have noted that the success of contingency planning is dependent on current planning and regular drills, simulations, or other testing. Prior to the attacks, executive branch COOP management by the National Security Council (NSC)[19] and FEMA and guidance for other executive branch agencies were all in place, and that guidance included requirements for agency-wide staff education as well as the testing and drilling of COOP plans. However, on September 11, 2001, some federal employees reportedly were unaware of these plans and some agencies found they had no way of accounting for, or communicating with, evacuated staff.[20]
- Budgetary constraints: The current budgetary environment is one of limited resources and of increased demand for a variety of homeland security protective measures, including organizational executive branch COOP planning. Acquisition of technology, infrastructure,

and supplies held in reserve for possible emergency may reduce the resources available for routine operations.[21]

- COOP plans not current: COOP plans are living documents, and any organization that implements a COOP plan must continue to refine and improve these plans to reflect lessons learned through both training exercises and real-world events. An organization's COOP plans are continuously revised and updated to confirm that essential functions can be maintained under all circumstances.[22]

These issues tend to reside in any planning process—COOP, BCP, and contingency—particularly as organizations move toward tactical levels in which scarce resources must be aligned as efficiently as possible to deliver the intended results. This leads to a situation in which, due to internal pressures, the bulk of the work begins to shift toward organizations positioned at the strategic level of the organization. Shifting away from the tactical and through the operational level leads to a situation where the tactical and operational level gradually refuse to accept the relevancy of the measures described, an issue that moves the organizations back toward the extremes of Quadrants 2 or 3 described earlier.[23]

9.9 Aligning with Preventive Safeguards

Preventive safeguards within the COOP context operate differently at different levels of the transportation system.

At the tactical level, preventative safeguards operate in terms of infrastructure assurances, which closely resemble BCP in its intent. The focus is on the ability to maintain acceptable minimum levels of performance while having plans and procedures put in place to move back to normal operating levels as quickly as possible. This incorporates a number of risk-based programs, particularly those that deal with the losses of personnel (occupational health and safety), assets (security), facilities (property management), information (information security), and operations (operations security and quality assurance). These programs all contribute in their own ways to ensure levels of performance feeding into operational levels in terms of both the expected quantity and quality of performance.

9.9.1 ABC Transport Example: Business Continuity Planning

ABC Transport has decided that the minimum acceptable level of performance for its distribution centers is 50%. Management at the distribution center has put in place a BCP and DRP to meet this requirement and to assure senior management that due diligence has been performed, These plans have been integrated into the day-to-day routines with drills and exercises further reinforcing the requirements. As a result, there is a high degree of confidence that the distribution center can maintain the minimum 50% under all but the direst of circumstances and that even in those circumstances it can move back to 50% quickly.

At the operational level, emphasis shifts form the model outlined herein, beginning with establishing redundancy and resiliency in the system such that disruptions are avoided or bypassed. The first focus of the operational level involves the establishment of networks necessary to identify evolving situations across the network. The needs of the operational level can be appropriately identified to suppliers if the company has made associations with those in the primary business lines and with those that provide key inputs. At the same time, operational levels seek commitments and agreements, preferably binding, providing availability of resources and priority of services to support response and recovery activities. Finally, operational levels need to track activities to have a real-time understanding of their status and a way to communicate this information to strategic levels as quickly as possible.

For example, ABC Transport's regional office has the single distribution center and five local offices. Trucks deliver packages from the distribution center to each of the local offices. Because the distribution center acts as a single point of failure, the regional manager supported the requirement for the distribution center to have a BCP and DRP. Each local office is simply required to maintain procedures to deal with the major risks identified in the threat and risk assessment.

The regional manager has identified several alternative routes and has decided to make certain that their vehicles are distributed somewhat so that a significant event will not result in the loss of all the vehicles. Trucks have been equipped with global positioning systems (GPSs) for road navigation, and the locations of the local offices have all been established so that drivers can navigate directly to any of the offices. As part of the service, ABC Transport has also decided to pay for a service that will automatically identify and route trucks around reported traffic disruptions. Finally, each office has been required to put in place a plan allowing for 48 hours of operations operating at 120% through the inclusion of additional workers. Should one office be disrupted, the two nearest offices would activate their plans, allowing the surplus capacity to be used to deal with packages that could be disrupted.

At the strategic level, COOP issues focus on the ability to develop the necessary communication networks, communities of interest and arrangements necessary to ensure both timely notification and the timely availability of resources. Communications networks are established to discover patterns associated with indications and warnings that an attack may be under way (post-Stage 3 in the U.S. Department of Defense model used to describe terrorist activities and particularly near Stage 5). They are also established to learn from successes and failures across the network. The next effort focuses on the communities of interest with particular attention being paid to making sure that the organization will receive priority treatment where interdependent on other organizations while also working within the context of industry associations to attempt to certify appropriate communications into the overseeing bodies. These arrangements focus on the ability to establish the best chances at detection while also promoting the ability to detect, respond to, and recover from events.

This effort focuses on abilities of the strategic headquarters to demand, through corporate policy, that the organization take into account redundancy and resiliency when looking at key processes. This operates at both tactical and operational levels, with the focus on the insistence that the organization, at operational levels, ensures that it does not tolerate single points of failure or become overly reliant on the capacity of any single node or conduit within the system.

ABC Transport's senior management has entered into a number of agreements. To assure the availability of energy, telecommunications, financial, and energy services (interdependencies), the senior management has entered into an agreement with the local emergency preparedness office. This has put ABC Transport's on the priority list for the local emergency preparedness organization, something that management markets in terms of its reliability as a carrier. Similarly, the company has undertaken to form an association of all delivery organizations, including labor interests, to make certain that regulators hear their voice. Finally, strategic management has set aside contingency funds that may be used to maintain operations at 140% operating costs for a period of one month in case of major disasters.

At the oversight level within the system, the preventive measures involve promulgation of regulations and similar directives based on results of strategic threat and risk assessments. The main effort is on coordination of regulatory development, not only with the stakeholders but also with other modes that traditionally intersect. This establishes that minimal gaps arise through the creation of disparate systems. At the same time, the overseeing layer within the system is in an ideal location to build necessary partnerships and learning systems so that the system might benefit from any lessons learned once they have been cleared of information that could expose specific vulnerabilities within the system.

9.10 Detection

Detection within the COOP context and within the transportation system relies on the creation of trusted and open networks based on communications and common interests. Effective detection operates both vertically and horizontally within the overall system, often across administrative organizations.

The key is to be able to communicate the specific indicators and warnings across the community of interest (tactically, operationally, and strategically) so that the community can begin to watch for those specific indicators. This would, given the likely sources of such information, begin at the strategic level and then be communicated into the operational and tactical level.[24] Having received such information, tactical and operational levels would feed such information back, allowing strategic levels to collect, collate, analyze, and assess the information to determine the extent to which those indicators and warnings exist in the system and determine whether or not mitigation strategies need to be activated.

This will not always be the case. In some cases, detection may involve patterns being recognized because of a number of independently reported and recorded events being passed on through the operational and strategic levels. This information, when collated and processed appropriately, forms a picture that becomes apparent at strategic levels when considered in the context of other information. Consequently, it is not enough simply to remain vigilant for specific indicators and warnings. Given the unpredictable and often creative nature of the adversary, any unusual event should be considered suspect until such a time as a rational and verifiable explanation for its occurrence can be determined (Figure 9.1).

This information flow operates within both the BCP and contingency planning processes, largely because the primary processors of the information reside outside of the immediate organization. In this context, the intelligence analysts are service providers to the organization, often not even residing in the organization. For the management team, this poses a significant challenge when looking at the issue of timeliness as it relates to the value of plans discussed earlier.

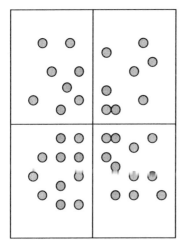 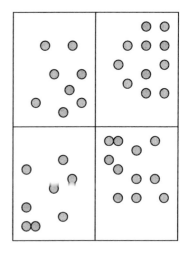

Figure 9.1 Identifying patterns needs whole pictures.

9.10.1 ABC Transport Example: Corporate Policy

Corporate policy requires managers to attend meetings with local authorities to determine if there are any threats to operations within the area. This information is shared between the layers to certify consistency and completeness. Additionally, every month the senior management's communications and operations arms publishes a password-protected and encrypted email for all managers that describes the number and kinds of incidents of suspicious behavior, loss, or damages suffered by the organization. Included in this report are quick synopses providing indicators and warnings derived from the investigations regarding those events.

Managers are required to brief their employees once per month on this issue to provide details of any suspicious behavior to report and so forth. Additionally, the company issues each employee a small, wallet-sized card providing five easy steps to reporting an incident that can be used to get them in touch with the regional security office. The reports collected by these organizations are fed back up the chain to the national level.

9.11 Response and Mitigation

Mitigation, in this context, involves the halting of the spread of contamination into the transportation system (a preventive measure) while also attempting to maintain at least some level of performance within the system. This may or may not be possible depending on the nature and severity of the attack or situation.

At the tactical level, mitigation is the identification of the scope of the event and then the execution of plans intended to contain the injury associated with that event. This generally involves a combination of security, contingency planning, and BCP. Security provides a workforce for this kind of event while also providing what is necessary to maintain the preventive posture during the period of increased impact. BCP, on the other hand, deals with identification of interdependencies as well as other inputs that lead to the successful completion of certain key processes. Additional levels of stability within the system are achieved by linking mitigation strategies with maintenance of minimum accepted levels of performance from a capacity management perspective.

9.11.1 ABC Transport Example: ABC Employees

Employees working at ABC Transport facilities attend, once per year, a half-day training and awareness seminar cosponsored by the company's safety and labor organization. During this seminar, employees are informed of the immediate actions that they should take should they discover an attempted attack is under way. By partnering with safety and labor, the critical infrastructure organization makes allies out of its potential rivals. Employee confidence in the workplace increases as their main representatives assist in ensuring that employees are aware of their responsibilities and rights.

The focus at operational levels shifts from an infrastructure assurance capability to a hybrid of infrastructure assurance and capacity management. On one hand, attempted efforts to maintaining levels of performance within the directly impacted area tend towards the infrastructure assurance. At the same time, the organization looks at the options that allow the system to reroute movement around the affected area while minimizing the impact on the mission. When the mission is impacted directly, injury is minimized through prioritization of movement making sure that the most critical movements take place.

Operational levels use hybrid approaches to approach a number of tasks intended to match assurances of infrastructure and maintenance of capacity. The first involves the release of information into preset and trusted communities to inform them of the disruption with the intent of maintaining the strategic-level capacity by allowing them additional time to assure their infrastructure. At the same time, the operational level is watching its own system determine capacity of various nodes and conduits. Once accurate or trusted information has been generated, the operational level is able to communicate that information to strategic levels so that impacts on the system can be determined.

9.11.2 ABC Transport Example: The Regional Office

The regional office maintains a list of key employees who are specifically trained to identify the extent of disruption and to generate options for reestablishing services. The purpose of this group is to ensure that any disruption does not spread to other parts of the company (containment). These employees are drawn from across the organization to provide a cross section of expertise and capabilities within the organization.

At the strategic level, mitigation and response involves the identification of macrolevel options to bypass the impacted area as much as possible. This functions similarly as spreading the load within the operational context, but with much greater impacts on interdependent systems (e.g., fuel, food). This may involve meeting or activating plans that include a number of levels of government, may be transsectored in nature, and may even involve the use of competitors under specified arrangements.

9.11.3 ABC Transport Example: Senior Management

Senior management, through its maintenance of current information in the ABC Transport capacity management plan, identifies situations in which the capacity in any organization drops below 70%. For these situations, the strategic level focuses on the nature of the disruption, its likelihood to spread to other parts of the organization, and its likelihood to persist for a significant period of time. This disruption looks at potential failures to complete current actions but also considers lag and delay caused by loss of capacity within the system. Management can support operational or tactical levels by assisting in generating the list of conditions to be met before the recovery of capacity begins.

As part of its training database, ABC Transport has the ability to search for people that possess special knowledge or skills useful on a temporary basis. The senior management can call on employees to provide expert guidance and advice on topics pertinent to controlling any disruption or containing any damage. These particular employees could receive a yearly bonus for being current on certain topics.

The first goal is to contain the injury associated with the event to guarantee that the contamination caused by the event is not allowed to permeate deeply into the transportation system. The second goal is to halt the injury of the contained event in such a way that minimally acceptable standards of operations are maintained to allow critical movements through the area. Finally, preset conditions for recovery operations are gradually worked toward until they are achieved.

9.12 Recovery

Recovery operations begin once the environment has been stabilized to a point that no further injury or contamination is taking place and that the likelihood that the situation will cascade beyond control. This concept applies across both COOP and BCP in that both approaches first halt the spread of the damage, then put in place the environment necessary to recover, and then begin the recovery back toward acceptable levels.

At the tactical level, this means that the action has been taken with regard to the direct impact and that the installation, or conduit, is beginning to move back toward normal operations. Gradually, controls around the impacted area are reduced, and efforts resume to return processes to their intended function. This will involve the reporting of results into the operational level, particularly in terms of estimated times when the organization will be fully functional or functioning at a preset level.

At ABC Transport's facilities, clearly identified conditions must be met before recovery takes place. After that, the tactical level will cordon off, contain, and gradually reduce the impacted areas until the company returns to normal operations. In addition to reestablishing the level of capacity, at the end of each day the tactical level sends a very brief synopsis to the regional management organization that defines the challenges, actions taken, work done, and risk remaining to the system.

At the operational level, recovery means returning the localized network to its normal state. The list of prioritized movements through the impacted area gradually includes more and more "less critical" movements until such a time as an acceptable level of performance has been achieved. Similarly, arrangements that may have involved other modes are slowly scaled back so that any disruptions caused during the mitigation process are reduced to their normal levels. The strategic level monitors this process so that a graceful adjustment of the system can take place.

Recovery operations for ABC Transport generally involve normalizing the local system. The regional office must slowly extricate itself from any temporary arrangements while also allowing those who have responded to the situation to withdraw little by little. The company takes care to maintain good lines of communication during this process while also monitoring for the presence of any suspicious or hostile activities.

At the strategic level, major adjustments had been made to divert pressure away from affected areas. These are gradually reversed until the situation returns to normal. This will likely be a timed approach driven by a combination of the timelines provided by the operational and tactical levels and the timelines associated with adjusting the movement within the system.

The role of the system-level oversight during this process generally involves the ability to establish controls to improve infrastructure assurance (increasing preventive measures or controls to prevent attacks against infrastructure) while bleeding pressure out of the system through the relaxation of controls. The relaxation of controls can expose the system to risk, leading to a requirement for specialized controls to be put in place to mitigate that risk.

9.13 Supply Chain Management Security

The ability to identify capacity within the system and potential disruptions in that capacity is critical to the role of supply chain management within any organization. The further ability to analyze alternative routes and eventually to predict natural disruptions works towards a more resilient supply chain, leading to a more robust sector.

Similarly, the various levels of prevention, detection, response, and recovery within the system are important to the supply chain security. The various levels (tactical, operational, and strategic) operate in much the same way, with tactical focusing on assurance and operational and strategic levels focusing more toward resiliency.

9.14 Questions

1. Is COOP limited to a government-based operation, or can it be applied enterprise-wide for the entire organization? If so, explain how. Based on your answer, explain how it would versus how it would not.
2. Describe in your own terms the interrelation among COOP planning, BCP, and contingency planning.
3. The culture in your organization shifts between wanting stability in its employment and less disruptions following accidents or similar events. Describe how BCP, COOP planning, and contingency planning can assist in meeting these desires.
4. In speaking with the management responsible for the supply chain, you discover that the company uses a just-in-time delivery system. Explain how, using the principles associated with COOP planning, BCP, and contingency planning and prevention, detection, response, and recovery you can identify appropriate levels of a just-in-case supply chain.

Notes

1. COOP is a BCP for federal government agencies. This kind of planning extends downward to state and local governments as well but usually does not affect private industrial sectors.
2. R. Eric Petersen, "Continuity of Operations (COOP) in the Executive Branch: Backgrounds and Issues for Congress" (CRS Report RL31857), April 21, 2003, http://www.au.af.mil/au/awc/awcgate/crs/rl31857.pdf.
3. Another term sometimes used to describe COG activities is enduring constitutional government (ECG). The terms appear to describe similar activities described in presidential national security documents as given in note 11. Petersen, "Continuity of Operations" (see note 2) does not discuss ECG or COG planning beyond any direct relationship to COOP planning. For a more comprehensive analysis of COG, see Harold C. Relyea, "Continuity of Government: Current Federal Arrangements and the Future" (CRS Report RS21089), January 7, 2005, http://www.fas.org/sgp/crs/RS21089.pdf.
4. Petersen, "Continuity of Operations" (see note 2).
5. A *cyber attack* is an incursion on a range of IT facilities and can range from the simple penetration of a system and the examination of it for the challenge, thrill, or interest to the entering of a system for revenge to steal information, to extort money, to cause deliberate localized harm to computers, or to cause damage to a much larger infrastructure, such as telecommunications facilities. Steven A. Hildreth, "Cyberwarfare" (CRS Report RL30735); Clay Wilson, "Information Warfare and Cyberwar: Capabilities and Related Policy Issues" (CRS Report RL31787).
6. See L. Elaine Halchin, "Federal Agency Emergency Preparedness and Dismissal of Employees" (CRS Report RL31739), August 8, 2003, http://www.law.umaryland.edu/marshall/crsreports/crsdocuments/RL31978_08082003.pdf.

7. The federal government produced several reports (and their findings) specifically relating to how COOP would be beneficial, government-wide, to all enterprise departments, organizations, agencies, and their entities. These reports are one of several Congressional Research Service (CRS) products related to government contingency planning and are periodically updated as events warrant. Petersen, "Congressional Continuity of Operations (COOP): An Overview of Concepts and Challenges" (CRS Report RL31594), January 14, 2003, http://www.fas.org/sgp/crs/RL31594.pdf addresses COOP planning in Congress. Halchin, "Federal Agency Emergency Preparedness" (see note 6) discusses pre-COOP activities relating to the safeguarding of federal personnel and evacuation of federal buildings. For a more comprehensive analysis of COG, see Relyea, Continuity of Government" (see note 3).

8. Although elements of COOP plans are available for some agencies, full plans detailing all potential responses are not public information, given their sensitive, contingent status.

9. The Y2K effort resulted because the format used to describe the date in electronic systems had been expressed in terms of two digits. When the year 2000 was encountered the computing system would crash because of confusion when computers misread "00" as being 1900 and not 2000. This problem led to significant efforts worldwide to check, correct, test, and monitor systems.

10. This is the premise that good strategic thinking would alleviate, if not almost complete eliminate, any risk inherent to the restoration of a given environment due to its plan. In most circumstances, a good business practice today can mean almost any good business practice of providing the continuity of an operation, regardless of whether or not it is complete.

11. Section 503 (1) of the Homeland Security Act of 2002 (P.L. 107-296) transfers FEMA, its responsibilities, assets, and liabilities to the Emergency Preparedness and Response Directorate of the new Department of Homeland Security. The transfer was effective March 1, 2003. For more information regarding the transfer, see Keith Bea, "Transfer of FEMA to the Department of Homeland Security: Issues for Congressional Oversight" (CRS Report RL31670), May 5, 2003, http://www.law.umaryland.edu/marshall/crsreports/crsdocuments/RL31490_05052003.pdf.

12. PDD 67, "Enduring Constitutional Government and Continuity of Government Operations (U), 21 October 1998," relates to enduring constitutional government, COOP planning, and COG operations. The purpose of enduring constitutional government (ECG), COG, and COOP is to ensure survival of a constitutional form of government and the continuity of essential federal functions. PDD 67 (http://www.fas.org/irp/offdocs/pdd/fpc-65.htm) replaced the George W. Bush Administration's NSD 69 "Enduring Constitutional Government" of 02 June 1992 (http://www.fas.org/irp/offdocs/usd/index.html), which in turn succeeded NSD 37 "Enduring Constitutional Government" of April 18, 1990 and NSDD 55 "Enduring National Leadership" of September 14, 1982 (http://www.fas.org/irp/offdocs/nsdd/index.html). The text of PDD-67 (http://www.fas.org/irp/offdocs/pdd/pdd-67.htm) has not been released, and there is no White House fact sheet summarizing its provisions.

13. In addition, Executive Order 12656 (Section 202; http://www.fas.org/irp/offdocs/pdd/eo1.htm) required that "the head of each Federal department and agency shall ensure the continuity of essential functions in any national security emergency by providing for: succession to office and emergency delegation of authority in accordance with applicable law; safekeeping of essential resources,

facilities, and records; and establishment of emergency operating capabilities." Among other things, PDD 67 required federal agencies to develop COOP plans for essential operations. In response to this directive, many federal agencies formed task forces of representatives from throughout the agency who were familiar with agency contingency plans. These agencies developed the COOP as a unifying concept that did not replace existing plans but, instead, superimposed COOP functions if a problem threatened serious disruption to agency operations. These plans identify those requirements necessary to support the primary function, such as emergency communications, establishing a chain of command, and delegation of authority. With the reduced threat of nuclear attack to the United States by the former Soviet Union and its successor nations, enduring constitutional government programs (the former COG programs) were scaled back in the early 1990s. Most of the resources of FEMA's National Preparedness Directorate were spent on ensuring the continuation of civilian government in the event of a nuclear war, through what are known as the enduring constitutional government programs. The directorate also supports ongoing studies through war gaming, computer modeling, and other methods. The April 1999 "Federal Response Plan" (FEMA 9230.1-PL) required the head of each federal department and agency to ensure the continuity of essential functions in any national security emergency by providing for (1) succession to office and emergency delegation of authority in accordance with applicable law; (2) safekeeping of essential resources, facilities, and records; and (3) establishment of emergency operating capabilities.

14. (http://www.fas.org/irp/offdocs/pdd/pdd-67.htm).
15. http://www.fema.gov/onsc/.
16. The following citations are as follows (specific Executive Order [EO] numbers, titles, dates, and URLs are provided afterward): The White House, *Federal Register,* Vol. 66, no. 246. EO 13241, "Providing an Order of Succession Within the Department of Agriculture," December 18, 2001, http://frwebgate.access.gpo.gov/cgi-bin/get-doc.cgi?dbname=2001_register&docid=fr21de01-153.pdf; EO 13242, "Provi-ding an Order of Succession Within the Department of Commerce," December 18, 2001, http://frwebgate.access.gpo.gov/cgi-bin/getdoc.cgi?dbname=2001_register&docid=fr21de01-154.pdf; EO 13243, "Providing an Order of Succession Within the Department of Housing and Urban Development," December 18, 2001, http://frwebgate.access.gpo.gov/cgi-bin/getdoc.cgi?dbname=2001_register&docid=fr21de01-155.pdf. EO 13244, "Providing an Order of Succession Within the Department of the Interior," December 21, 2001, http://frwebgate.access.gpo.gov/cgi-bin/getdoc.cgi?dbname=2001_register&docid=fr21de01-156.pdf; EO 13245, "Providing an Order of Succession Within the Department of Labor," December 21, 2001, http://frwebgate.access.gpo.gov/cgi-bin/getdoc.cgi?dbname=2001_register&docid=fr21de01-157.pdf; EO 13246, "Providing an Order of Succession Within the Department of the Treasury," December 21, 2001, http://frwebgate.access.gpo.gov/cgi-bin/getdoc.cgi?dbname=2001_register&docid=fr21de01-158.pdf; EO 13247, "Providing an Order of Succession Within the Department of Veterans Affairs," December 21, 2001, http://frwebgate.access.gpo.gov/cgi-bin/getdoc.cgi?dbname=2001_register&docid=fr21de01-159.pdf. The following citations are as follows (specific EO numbers, titles, dates, and URLs are provided afterward): The White House, *Federal Register,* Vol. 67, no. 8. EO 13250, "Providing an Order of Succession Within the Department of Health and Human Services," January 11,

2002, http://frwebgate.access.gpo.gov/cgi-bin/getdoc.cgi?dbname=2002_register &docid=fr11ja02-126.pdf; EO 13251, "Providing an Order of Succession Within the Department of State," January 11, 2002, http://frwebgate.access.gpo.gov/ cgi-bin/getdoc.cgi?dbname=2002_register&docid=fr11ja02-127.pdf. The White House, *Federal Register*, Vol. 67, no. 55, EO 13261, "Providing an Order of Succession in the Environmental Protection Agency and Amending Certain Orders on Succession," March 21, 2002, http://frwebgate.access.gpo.gov/cgi-bin/getdoc. cgi?dbname=2002_register&docid=fr21mr02-105.pdf. There does not appear to be a similar executive order specifying succession in the Department of Homeland Security.

17. See U.S. Department of Energy National Nuclear Security Administration, NAP-2, "Establishment of Line of Succession for the Administrator, National Nuclear Security Administration," May 21, 2002, http://www.nnsa.doe.gov/docs/NAP-2.htm; U.S. Department of Transportation Federal Highway Administration, FHWA Order M 1100.1a, FHWA Delegations and Organization Manual, Part I, Chapter 2, "Order of Succession," September 30, 2002, http://www.fhwa.dot. gov/legsregs/directives/orders/m1100.1a/doa_ch02.htm.

18. Federal Preparedness Circular 65, Federal Management Agency (www.fema.gov/ppt/ government/coop/vital_records.ppt) http://www.fema.gov/onsc/docs/fpc_65.pdf.

19. The National Security Council was established by the National Security Act of 1947 (PL 235 - 61 Stat. 496; U.S.C. 402), amended by the National Security Act Amendments of 1949 (63 Stat. 579; 50 U.S.C. 401 et seq.). Later in 1949, as part of the Reorganization Plan, the council was placed in the Executive Office of the President. Since its inception under President Harry S. Truman, the National Security Council is the president's principal forum for considering national security and foreign policy matters with his senior national security advisors and cabinet officials. The council also serves as the president's principal arm for coordinating these policies among various government agencies. http://www.whitehouse.gov/nsc/.

20. Petersen, "Continuity of Operations."

21. ibid.

22. Office of National Continuity Programs (www.fema.gov/government/coop/ index.shtm)

23. At the tactical level, the trend tends to move toward a "performance sacred, protection sacrificed" posture driving a perception that risk is not addressed appropriately. This perception generally focuses on a relatively narrow area with the result being that the overseeing body is eventually compelled to dictate the specific requirement, often in the form of a contingency plan. Though the end goals would generally be beneficial, this cycle is both disruptive and potentially destructive as the requirement, applied generally, may only be appropriate in localized or regional contexts.

24. Using a linear approach to this issue will generally lead to confusion. Though the threat and risk assessment modelling process looks at the kinds of threats working toward issues that are more specific, this does not preclude any ability to look at specific events to determine whether an indicator or warning exists. For the practitioner, understanding the information processing cycle (or even an intelligence cycle) becomes very valuable in terms of being able to understand where information resides and how it is gradually processed.

Networks and Communities of Trust

10

Objective: At the end of this chapter, the reader will understand the following:

- The definition and implementation of a trusted network and how it is utilized within the transportation system
- Fundamental concepts surrounding community and why trusted networks are vital and important to both the community and the organization serving it
- The role of memoranda of understanding, mutual aid agreements, and other binding agreements
- How to encourage volunteering for the trusted community

10.1 Introduction

The transportation system operates not as a single entity but as a collection of communities in varying states of competition. The creation of trusted networks of nodes and conduits uses common criteria through the process of certification, accreditation, and registration and licensing facilitates capacity management. Certain variations of risks, safeguards, and vulnerabilities are allowed to operate within the system. Although risk can never be totally removed from the system, the fact is that when risk is understood it becomes more manageable across tactical, operational, and strategic levels.

The same holds true for infrastructure assurance aspects of protecting the transportation system. Again, the key here involves reducing the unknown. Only in this case can we focus in on understanding of how entities respond under certain conditions. This requires a different level of networking on the part of the various levels of participants.

171

10.2 Value of Community Involvement

Many communities grew up along the corridors or nearby nodes within the transportation system. Significant cultural ties exist between the community and the actual infrastructures themselves. For those involved with the infrastructure assurance process, this provides a significant opportunity to build communities involved with the prevention of, detection of, response to, and recovery from events within the system or (more importantly) to reinforce resiliency levels in local systems.[1]

The *community of interest* consists of those operating within the transportation system (often referred to as *participants*) who are impacted by system events or who provide goods or services within and throughout the transportation system. The critical infrastructure protection (CIP) security professional may wish to expand this further and include those who feel that they have an economic, social, or similar stake in the services provided by the sector. In essence, the CIP security professional should consider including all participants of the community even though there may be limitations assumed based on their level of involvement.

Within the context of infrastructure assurance (for either counterterrorism or anticrime requirements), being involved provides a significant advantage. Having full involvement from the community denies a potential attacker a community within which he or she can conceal activities, analyze, or plan while in safety or can gather the materials necessary to conduct the attack. This is due to the community's involvement in the cycle of prevention, detection, response, and recovery.

10.3 Prevention

Communities tend to rally around and protect those aspects that are important to them. Where the transportation system has played a significant role in the evolution of the community, establishing and reinforcing that tie can be invaluable for the CIP security professional. The goal is to solicit involvement from outside volunteers to support the interest of the organization at any of the tactical, operational, or strategic levels. This kind of effort provides value because it overcomes one of the most significant challenges involved in protecting infrastructure—the concept of ownership of infrastructure and a stake in its assurance.

Prevention denies potential attackers an environment within which they can plan, coordinate, and carry out an attack with impunity. The attacker will attempt to exploit any perceived conflict within the organization, perhaps by concealing his or her motives to trick the community into supporting his or her activities. Removing the community base from the attacker leaves him or her less able to carry out plans.

10.4 Detection

Detection involves being able to recognize suspicious activities or conditions and reporting those conditions to the right communities or positions within the organization. Although technological advancements have received much public attention in this respect, the security community has come to realize that the individuals who have taken ownership of the security challenge and have amassed experience within that environment provide a significant advantage. This is largely due to familiarity of those individuals with operations and equipment used in those areas.

The detection of suspicious behavior requires individuals to report that behavior. This can pose a significant challenge, especially if the individuals serve as the only means of detection within the organization. Though it seems logical enough most individuals would report suspicious behavior as part of their job security or if community safety required it, not all persons have this point of view. Within the transportation system, this has led to a situation in which individuals must undergo background checks. These checks provide a valuable tool if they are supported by the appropriate policy decisions within an organization, but they can also be just a snapshot in time. As a result, CIP and security organizations will want to ensure that this is not the only means of detection.

By building communities of interest and casting the nets wide, the CIP security professional has the opportunity to communicate to the community what are generally referred to as *indicators*, requesting that any individual who notices an indicator contact the appropriate part of the organization upon its discovery. This requires that the CIP security professional clearly identify what the actual indicators are, have a way to screen out false positives and negatives, and have some method of recording reports of behaviors for future trend analysis.

Within this context, communication plays a vital role. The CIP security professional should not limit those lines of communication to the traditional infrastructure. Instead, redundancy and resiliency can be built into the system, particularly where communities can be brought on board, through approaching communities that have already established alternative communications networks. Consider, for example, a power failure along with the disruption of the land-based communications system, such as a telephone system. The 2003 blackout in Ontario, Canada, and much of the eastern coast of the United States, disrupted cellular towers due to the volume of calls being pushed into the system. In this case, communities such as amateur radio operators could be invaluable in the passage of information to identify that a disruption has taken place. By establishing links into these kinds of communities, the CIP security professional, along with emergency response organizations, can begin to generate a clearer view of the impact—a key factor in being able to respond appropriately.

10.5 Response

One of the key aspects to critical infrastructure protection involves the ability to halt the spread of contamination into the overall transportation system. Here again, the concept of community building plays a significant role, particularly with respect to the role of first responders.

One of the key challenges when dealing with multijurisdictional organizations is that those organizations often work using different standing operating procedures (SOPs) or from different contexts. Discovering this during a crisis or emergency is not ideal. Given that the transportation system relies on services drawn from national, regional (state or provincial), or even municipal (including tribal) communities, each node and conduit that would normally call on these should, at the earliest opportunity, start to begin building lines of communication and community.

For the first responder, the key benefit to this involves an in-depth knowledge of the environment. Understanding terminology, geography, lines of communication, organizational structures, and similar characteristics can prove invaluable to those seeking to apply solutions or to generate answers with all possible speed and under difficult conditions.

For the CIP security professional, there is great value in understanding resources called on to assist in response to an event. By being able to identify potential vulnerabilities within the response (e.g., disruption of a vessel's delicate balance by flooding it during firefighting) to those with the right kind of expertise, the various kinds of responses proposed kinds of situations can be tailored to mitigate the risks.

To accomplish this, the CIP security professional, along with management at the various levels of the organization, must begin building communities early, generally at the same rate as the community-building efforts during the preventive phase or as early as possible in the process. Ideally, first responders should be involved with the generation of the response plans, given that they are the experts in their fields.

While this community is built, the CIP security professional should be particularly mindful of any interoperability issues, particularly with respect to communications. Hardware, software, and cryptographic incompatibilities can lead to serious challenges that must be identified during the testing phases of the activity.

10.6 Recovery

Recovery will place significant pressures on the community, particularly where impacts have spread into the community at large. In this case, the ability to recover will be directly linked to the community's perception of the importance of the transportation system. Where the infrastructure is considered to be of primary importance, the management of the organization may stand to realize significant benefits from the various levels of government in assisting in restoring services associated with that infrastructure. If the transportation system is not incorporated within the plans of the local or regional community, management may find itself competing for scarce resources—often at a premium—since the majority of the work has been prioritized away from the organization.

10.7 Community Building as a Continuum

The community-building approach is an ongoing process that involves continuous efforts within each of its phases. This is largely due to the nature of resources being considered and how the planning cycle functions.

The first effort of community building identifies varying kinds of responses available within the area and which organizations control those resources. This is particularly pertinent when dealing with police, fire, and medical organizations working within the community. Consider similar circumstances involving pandemic planning, where the best value for detection may reside within local hospitals and clinics that are equipped at identifying and handling these kinds of cases. Thus, a clear line of communication to the correct established expert will provide information that something is occurring. These organizations may point toward certain centers that collate that information, such as the Centers for Disease Control and Prevention in the United States, which handles communications for such events. Understanding reasons as to why certain lines of communication need to be maintained, and often compromising somewhat, is important in being seen as a member of a community that can be trusted not to interfere in their own response during a difficult situation.

The second effort involves a level of learning across organizations. This may be as informal as awareness sessions or visits or can be as formalized as cross-training of individuals and holding exchanges. The key here is to generate internal lines of expertise that can begin to spread key requirements of both efforts of the community systems throughout each other. For example, allowing a member of a port security team to work with a volunteer firefighting department permits the visitor to remain current within the field and to bring knowledge back into the port authority. That individual can communicate the requirements of the port authority more effectively to the firefighting community. This would help in developing appropriate responses.

The third effort involves consistent interaction between various communities through the implementation and activation of drills and exercises. This allows individuals within the organization not only to understand the nature of the response happening around them, but also to begin to build a level of trust between the organizations. This leads to two requirements. The regime preceding this exercise should be suitably valuable and challenging to individuals who are used to taking significant levels of risk. Team building requires this kind of approach as it develops informal bonds. The second involves retaining those individuals that have become part of the team. In this respect, the CIP security professional may consider looking at establishing methods by which individuals can remain involved with the team, although traditional and formal ties may have been severed. For problem-solving purposes, the breadth of knowledge that can be applied through this kind of system can lead to a reduction in the risk associated with failing to identify vulnerabilities.

10.8 Setting of Arrangements

For the CIP security professional working within the transportation system, the key is to ensure that appropriate arrangements and lines of communication are identified, established, exercised, and renewed. Linear approaches to this issue generally face significant issues, so the CIP security professional will want to include mechanisms for the review, resubmission, and reinforcement of commitments made between the two parties. In some cases, arrangements may be informal, such as the identification of communities such as the local amateur radio operators or other volunteer organizations. In other cases, requirements may be more formal, such as those between different levels of government or even national entities. These are often coordinated through such mechanisms as memorandums of understanding (MOU)[2] between volunteer organizations (see Appendix A involving the Canadian Red Cross and Radio Amateur Canada for an example of an MOU).

The key to these arrangements involves establishing an agreement that clearly communicates needs, expectations, and obligations for each party. In many jurisdictions, the tools for this have been identified and established within a legislative framework. In other cases, MOUs will more often suffice, as long as they are of sufficient strength that both parties can trust them. This can generally be identified in terms of the following:

- Identifying conditions under which the agreement comes into force
- Identifying specific parties involved
- Identifying roles of each party, including values provided to the system
- Identifying specific goods or services to be provided, including the setting of a service standard with relation to an agreed on criteria
- Understanding how and under what conditions the agreement can be terminated and how to communicate that termination

One key to building this level of community involves the recognition of contributions, particularly where informal groups are involved. Consider that many of the individuals are putting the community ahead of their own personal interests. For the amateur radio operator, this may involve a willingness to remain behind with meager infrastructure during difficult periods to ensure that accurate information to or from the community caught in the event can be communicated. Persons have put the good of the community or society ahead of their own. In some cases, their very lives have been at risk, especially during a natural disaster or severe weather event.

10.9 Communities and Council Building

A requirement in any community is the ability to communicate information and to trust that it will be used appropriately. This can be a challenge in cases in which competitors or political interests are involved, as individuals of influence may not want information shared between various parties. In some cases, this information may be sanitized or presented into a larger forum through councils, such as those set up in terms of industrial groups or associations. These associations or councils serve as filters that can collate, clarify, and sanitize information where appropriate to prevent its use outside of its intended purpose.

Similarly, investigative and emergency response organizations often form councils that serve to share information of interest to each other. For those responsible for security and emergency preparedness within the transportation system, great care should be taken to attend these meetings, communicating results to management. This is done for two reasons. The first is that it keeps management aware of any new situations within the environment. The second is that it identifies whether or not those organizations are addressing any of the company's general interests. If not, perhaps management would want to reconsider participation in those particular communities.

10.10 Tactical, Operational, and Strategic Considerations

Levels of community building do not simply exist at one level but also across all levels of the system. This is essential if the system is going to function appropriately. Management needs to understand the nature of the information that is both received (risk information) and sent (risk, particularly vulnerability, information) if it is to maintain its primary objective of preserving the organization. This can pose a number of challenges.

At the same time, each layer of the organization will require a level of reassurance that its participation is in support of organizational goals. Although part of this can be determined from mission statements and their alignments, there are human elements that cannot be discounted, such as the reluctance of an individual to harm the community to which he or she has made a commitment. Thus, both corporate policy and senior management support will have to be constantly reinforced, particularly during periods of difficultly, so that tactical and operational levels are reassured in their own activities.

10.11 Communities, Trusted Networks, and Operations

The concept of the trusted network refers to the reality that the transportation system operates as a collection of entities in various states of competition. The certification and accreditation aspects of the trusted network address the issue of whether or not partners possess and maintain adequate levels of infrastructure assurance and capacity management capabilities. The registration and licensing allows the management to assure its clients that their goods or people will be delivered expeditiously and safely. By establishing trusted networks across all four activities (prevention, detection, response, and recovery), the various nodes and conduits controlled by those entities (which maintain the appropriate level of controls) are added into the corporate network as a second tier that can be trusted to deliver on its promises.

This requires a number of subagreements to be in place. The first, and most obvious, is the maintenance of controls along the lines of infrastructure assurance and capacity management. The second becomes the agreement process by which either party—under certain conditions—can reopen discussions to reset terms. The final involves the business-level commitments that, having established the trusted network, indicates the reward for being part of that community (e.g., preferential treatment in the contracting processes, new supply lines). What will gradually form if this kind of system spreads is essentially a chain of agreements that will bind organizations together across the various modes and administrative areas and will become identified as a trusted network that is validated through monitoring within the community.

To accomplish this there has to be an understanding of each other's systems and the ability to communicate between them. At this point, the coordination challenge becomes very apparent at the hardware, firmware, software, data categorization, and procedural levels. Within this trusted community, the ability to communicate is crucial as it represents a narrower community that may become more vulnerable to a disruption within itself (where the community is bound only to the trusted community's nodes and conduits) or less able to evolve (as the passage of information is nonexistent, incomplete, or inaccurate). Addressing communications at the field, office, and corporate levels becomes a key component within the trusted network's ability to function.

10.12 ABC Transport Example

The company has determined that it needs to generate additional information pertinent to threats and risks against its operations. Concurrently, the company wants to launch an advertising campaign focusing on its ability to assure its clients of on-time delivery anywhere in its operating environment. Thus, management has made a number of decisions to position the company in such a way that there is a high degree of confidence that it will receive the kind of support it needs from first responders should anything go wrong. Management has also decided that it will expand its understanding of capacity available for use in the market. Concluding both of these efforts and normalizing them within corporate processes will give ABC Transport a competitive edge of some of its rivals.

The first policy decision directed operational and tactical levels of the organization to approach the local emergency management office, offering services in case of emergency while also identifying the capacity of the organization. In this case, the local offices and distribution nodes, anticipated to be disrupted, could be used to assist in the distribution of aid while a limited number of vehicles would be made available to assist in the movement of supplies.

As a result, the local emergency management office added ABC Transport to its list of potential contributors to its response to an emergency. In addition to any political points scored by supporting the various communities in which it was operating, participation allowed ABC Transport's security organization to develop stronger lines of communication and alliances with various government departments and agencies. These could provide threat information so that it could maintain good preventive and infrastructure assurance postures. This was repeated at the operational level.

At the local level, managers ensured that the lines of communication with local emergency responders were normalized and that both parties understood their mutually supportive roles in the case of an emergency (defined in mutual aid agreements). This assisted in developing closer lines of communications with medical, police, and fire organizations that were able to better and more clearly define the company's needs when responding, to a large or area-wide incident. This would lead to a more appropriate response. At the same time, a number of joint exercises served to build better understanding and cooperation across the various communities.

At the operational level, two awareness campaigns were launched. On one hand, police assisted ABC Transport in a campaign that focused on identifying suspicious behavior and surveillance. Employees, clients, and drivers were strongly encouraged to attend these briefings. In return, ABC Transport assisted the police with their suspicious package handling awareness campaigns, making certain that the credits on the awareness videos included the company name and logo to show its participation in the community. These awareness presentations were provided to employees and managers alike under the guise of environmental awareness and of providing a safe, secure, and comfortable working environment.

Close lines of communication were established with medical and fire communities. This allowed ABC Transport to identify specific areas of concern within the medical community with respect to the ability to move vaccines and other medical goods quickly and securely. ABC Transport was then able to equip a small number of its transport vehicles, such as vans and several panel trucks, with necessary equipment to move these materials safely while opening up a potential new business line in moving medical supplies between hospitals. Additionally, vaccinations and similar considerations were offered to the ABC Transport drivers and planners coordinating this service as part of the medical community's plans.

The fire department revealed that they had concerns about movement of dangerous goods by couriers and similar companies. ABC Transport provided its contact information and a guest user account to the truck loading system to the local emergency dispatch center in the area so that first responders would always be able to access the manifest list for any given vehicle. Accessing data within this account sent a notification to the regional operations personnel, who were then able to use that common point of contact, under arrangement (usually asigned document), to determine the specific nature of the emergency with the vehicle.

ABC Transport encouraged operational-level managers to reach out to key suppliers, providing them with an opportunity to enter into a mutual agreement. This agreement defined that, if certain measures were put into place, preferential treatment would be possible on upcoming contracts.

Within these agreements, a number of conditions was laid out. This included providing point of contact information; refining key communications terms and procedures to facilitate the clear, concise, and timely passage of information; and providing for periodic inspections of facilities to verify that terms and conditions were being observed. As a result of these agreements, the company was able to refine its plans and to chart additional routes for the movement of parcels should there be a disruption in any of its own routes.

10.13 Questions

1. Explain how an MOU is different from that of a legal document (e.g., a contract or user agreement).
2. Using the concept of nodes and conduits discussed earlier in this book, and the concepts of prevention, detection, response, and recovery,

describe how the establishment of a trusted network with appropriate controls could enhance the ability to choose additional nodes and conduits. Consider how this would affect dealing with single points of failure.

3. Provide an example in which the participants of a trusted network could hamper operations. Provide an example in which participants of a trusted network could enhance operations.

4. What advantages and disadvantages can be realized by using volunteers during difficult situations? What leadership and corporate culture challenges do you foresee with this approach?

Notes

1. The concepts of prevention, detection, response, and recovery parallel another approach that focuses on deterrence, delay, detection, denial, and detention used across parts of the security industry. The latter is particularly prevalent within the American Society for Industrial Security (ASIS).

2. An MOU is a legal document describing a bilateral agreement between two or more involved parties. It expresses a convergence of will between the listed or outlined parties, indicates an intended common line of action rather than legal commitments, and is a more formal alternative to a gentlemen's agreement but generally lacks the legally binding power of a contract.

Establishing and Monitoring Learning Systems **11**

Objective: At the end of this chapter, the reader will understand the following:

- How information sharing will play an even increased vital role within and throughout the transportation system
- Current methodological and legal issues impeding the sharing of information
- Methods for implementing response and mitigation and recovery of the system if it is degraded or compromised

11.1 Introduction

Evolution, in the words of Charles Darwin, involves the survival of the species most suited to its environment. Though his origin of the species focused on biological entities, the same may be said for the systems that protect a critical infrastructure. Changing operations, integration of new technologies, and streamlining of processes along with the rotation of personnel within the organization all affect the processes that drive the mission forward. As a result, they have an effect on the criticality of certain persons, objects, facilities, information, and activities. The changing geopolitical world, arguably more fluid now than during the cold war, leads to an evolving threat environment where various threat agents are not content to simply wait for the environment to change but also will attempt to identify and exploit vulnerabilities to accomplish their aims. Changing market pressures, ranging from regulations to competition, lead to the need to adapt mitigation strategies and procedures to realign with their environment. In essence, a critical infrastructure protection (CIP) program that does not integrate a learning system is bound to be effective for a limited period but will, if remaining stagnant, expose the organization to an increasing level of risk as time progresses. Like any security- or risk-based program, as people change so do systems and how they integrate with each other. The learning and awareness processes must be flexible and responsive to sociological, political, and economical changes; otherwise, all of this will be for naught. To quote Albert Einstein, "The only source of knowledge is experience."

11.2 Intent of the Learning System

Learning systems as discussed here have three goals. The first goal involves ensuring that unnecessary loss or waste from failure will only happen once. The second entails the ability to identify what is working adequately. The third goal encompasses the learning system, as it should be capable of communicating success or innovation across systems to maximize potential benefits.

The goals of the learning system apply to tactical, operational, and strategic levels of an organization. What will shift are the topics of interest to each level. At tactical levels, infrastructure assurance figures highly, including resiliency and redundancy factors associated with the ability to meet its promises. At operational levels, the need is the ability to identify alternatives and to maintain levels of performance both internally and, depending on any agreements in place, within the trusted network. For the strategic level, the ability to identify, respond to, and eventually predict change in the environment is a key benefit. At all levels, the intent of learning systems falls within the general category of waste prevention in that it is one more tool used to minimize risks associated with both realized loss and unrealized potential. Finally, at the system level the learning system provides opportunity for capacity management in support of national and sometimes transnational goals.

11.3 How the Intent Is Met

Generally, management cycles will incorporate some level of analysis, planning, implementation, and review capability against several criteria for success. Learning systems follow similar principles with a set of common features, which include the following:

- Actions are assessed or evaluated against specific criteria sets for success.
- The divergence between success and observed results is identified and then evaluated to determine what caused the divergence.
- Key factors contributing to success and failure are identified and linked.

- Information is sanitized in such a way that vulnerabilities are not exposed beyond communities of trust.
- Feedback is sought regarding other alternatives and is collected into a repository of information for future reference.

11.4 Assessing or Evaluating against Criteria

Remember that concepts of infrastructure assurance and capacity management are both risk based, and that management decisions, about the allocation of resources, are made based on how best to deal with risk. This makes risk the natural criteria to use when looking at learning systems. Though performance metrics associated with risk may provide some gradual reduction of likelihood or impact, the key to determining whether the system is performing as intended is really an evaluation as to whether the predicted risk in the system is in balance with the observed risk within the system.

In Figure 11.1, an event has taken place, and the risk it poses to the organization was higher than anticipated. This could be the result of a failure to understand or include elements within the impact or a failure to assess correctly the probability of the event. The location (indicated by the star) rests below the diagonal line.

This approach provides a validation tool for the threat and risk assessments. Thus, it has two core values. The first value involves validating processes used

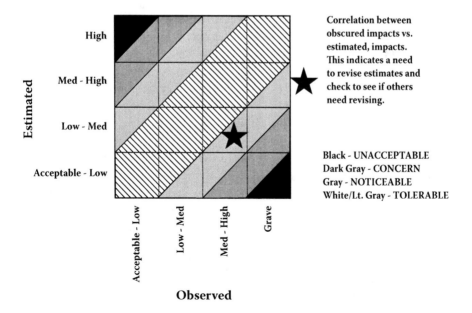

Figure 11.1 Observed vs. estimated validation.

in that process to support its future arguments or identifying areas that may require refinement to become more accurate. Second, it provides a natural focus on being able to assign measurable values into the threat and risk assessment system. Although it is unlikely that a fully credible and measurable system will evolve in the near future, the integration of at least basic measurement requirements leads itself to more predictable and repeatable results. This approach may be used at any of the tactical, operational, and strategic levels as long as it remains tied to threat and risk assessments at their appropriate levels.

11.5 Prioritizing Based on Divergence

As the divergence between actual and estimated risk increases, the gap between management expectations and delivered results also increases. For the CIP security professional, being able to identify and then to demonstrate issues have been addressed can be critical in ensuring the program. Given constraints in resources, the CIP security professional may wish to have management identify thresholds—work such that events fall within the anticipated boundaries that describe these thresholds.

11.6 Determining Causes

Where divergence occurs, the key is to determine why divergence occurred to maintain credibility of the system (insofar as a matter of the integrity of the system when referring to the C-I-A triad).[1] This could be the result of a number of factors, each of which must be considered.

One question involves checking initial information received to determine if, in fact, an error was made in the initial calculation of information. This process parallels quality assurance program requirements, requiring only a return to the original threat and risk assessment to determine whether the information was in fact treated correctly.

Having determined that the information was handled correctly, it is now important to determine whether the data entered into the system were, in fact, accurate. Where threat information itself is concerned, this is a matter of simply validating information provided by the appropriate lead agencies. Most of this validation is likely to occur by evaluating facts associated with the event against the impacts, probability, or likelihood as they were assessed. For example, it may have been estimated that the loss of an asset would have a relatively inconsequential impact. This consideration applies to criticality, probability and likelihood, risk, and vulnerabilities.

If the information entered into the system has proven valid, the next step involves determining whether the mitigation strategy and its specific measures performed as anticipated. This is where a well-defined CIP plan provides significant value to the organization as it provides the CIP security professional a detailed list of things to verify. Where the divergence occurs because the safeguards assumed within the plan were not functioning as intended, the CIP security professional has identified a new vulnerability in the system that should be addressed.

Once the specific issue has been dealt with, steps must be taken to address the vulnerability, including the measurement criteria that can be used to determine whether the measures were in fact successful.

11.7 Communicating Results

Whereas the first step addresses the vulnerability, the second tries to maximize the benefits associated with the information. In this specific example, the goal is to protect the organization against similar kinds of failures. There must be the capability to communicate information in such a way that it provides value to the organization but at the same time does not provide any significance or value to hostile parties.

Within the organization, communication of this information is straightforward. The results of the process and the follow-up action taken can be communicated up throughout the management chain and recorded into appropriately protected repositories so that it can be of benefit throughout the organization. This is largely because senior management, as the holder of these information repositories, has the overriding stake in its use to promote the success of the organization.

Another issue involves the trusted network that is bound to certain kinds of behavior as a condition of remaining part of that network. This generally involves a mutual agreement among all interested parties to certify that any sensitive information is handled appropriately. The challenge is to identify the entity responsible for maintaining and securing information repositories. This role usually falls to the state or a similar body that operates outside of the business interests of the various participants within the system.

The challenge lies with the sharing of information that may or may not identify vulnerabilities in the system. Though concepts such as information sharing and assessment centers (ISACs)[2] are intended to provide a solution for this, there is a significant challenge involved in trying to get any company to share information when it feels that its competitors could gain access to that information. At the same time, industries that are heavily regulated or are subject to penalties under regulations are very unlikely to share information with the regulator when they feel that such sharing of information could result in prosecution.

For the regulator, the solution is relatively simple. The philosophy, expressed in departmental policy, might be to ensure that a partnering approach is taken to ensure the maximum possible buy-in when complex issues are being discussed. This approach has been used with some success in both Canada and the United States.

For industry though, the challenge is somewhat different. In this case, the concept of sanitization plays a significant role. This means that information being put into the system must appear to be anonymous (except to the holder of the information) and that the repository of the information must be protected so that it does not become a source of data for those involved in industrial espionage. Included in this is a broad corporate security effort making certain that all aspects of the information are protected—from its creation to its destruction.

11.8 Challenges with ISACs

Notwithstanding whatever business model has been chosen or is used, an ISAC may eventually face legal, technical, and socioeconomic challenges that might affect its operations and ability to disseminate and distribute threat analysis information accurately and in a timely fashion. A 9/11 Commission member stated that current ISACs do not work and that his inclination was either to reform the ISACs correctly or to discontinue their use entirely.[3]

Of these challenges, probably the most significant factor is the human factor, which includes one of the following:

- Information technology staff who develop, deploy, and administer security technologies of detection, prevention, and response to attacks have to be not only subject-matter experts of varying degrees but also trustworthy and loyal.
- Security systems may be influenced by a variety of human and organizational factors, such as training and end-user awareness issues, information availability, and their policies.
- Applied security technologies, devices, and their systems might conflict with socioeconomic or political values.
- Funding (or lack of it) poses the greatest challenge since ISACs require personnel.

Other challenges associated with operating and maintaining an ISAC pertain to the administrative and legal aspects, some of which continue to plague both private- and public-sector organizations. There are administrative concerns with the operation of an ISAC: Essentially, who is going to maintain and administer the ISAC, how will it be run, and how will it notify organizations if an incident or event occurs? The applicable framework depends on how it is incorporated, who is involved, and what is involved. For example, an ISAC may be considered a partnership or consortium of several interested agencies, departments, or corporate organizations.

It may be a viable option to establish the ISAC as a limited liability corporation; however, if established as a for-profit business it may be conceivable that a public company with shareholders or other corporate investors may be involved. It may also be possible to consider a nonprofit business model instead, such as an institution or consortium, with a mandate to foster on-line trust and confidence through the following:

- Efficient, effective, and timely dissemination of information about any threats, risks, or potential attacks as well as reported incidents or events that are relevant to the sector being monitored
- Preparation of short, medium, and long-term advisories about current and future speculations of both current and potential challenges, threats, and threat scenarios against the specific infrastructure (and its sector) as well as information about specific incidents or events

Antitrust concerns also need to be addressed, as it is not desirable to have direct involvement between competing private-sector organizations. Corporate secrets are viewed and treated as assets of the participating corporation.

Thus, the dilemma is whether to share, disseminate, or distribute information to any potentially competing corporations within the same or similar industry or simply to do nothing.[4] Assuming that the primary function of the consortium is to foster the monitoring and detection of threats and to issue an alert based from an early warning, information sharing, and collaboration, one may speculate that it should not face anticompetitive restrictions from the consortium or institution.

One method of exchanging information might be through a series or set of simple exchanges of information (statistical information, marketing research conducted from outside research groups, or comparative studies conducted against either the corporation specifically or the industry in general). More importantly, it should not involve restrictions of any actions or undertakings (or, for that matter, any recommendations provided or given by the participating member organization) that would potentially induce other participating member organizations to behave and operate in an identical or similar manner.[5]

11.9 How Would Information Be Shared?

Many governments (at a national level) usually wish to undertake research and development projects based on any data or information that have been collected; however, the consortium might allow competing private-sector organizations or corporations to cooperate, providing that the following conditions have been met:

- Conduct joint research and development of products, services, processes, and methodologies with the notion that everyone benefits from its outcome
- Conduct joint exploration research programs that produce results from research and development carried out jointly from a prior nondisclosure agreement between all interested and participating member organizations
- Conduct joint research and development of products, services, processes, and methodologies that may exclude any potential joint exploitation from any conducted joint research or development programs and their results

The fulfillment of these requirements are, nonetheless, not always feasible and in some circumstances are not always possible. Not all corporations or organizations are willing to cooperate within this level of capacity[6] as this form of highly corporate competitiveness often leads to intellectual property legal embroilments and lengthy lawsuits.

11.10 Legal Issues with ISACs

Until now, many assumptions have been made that the establishment of a private entity consisting of companies, public agencies, departments, and institutions would have access of payment in some form of remediation or compensation. This can pose a significant challenge when looking at the administration of public funds. Simply providing those funds without some level of control over the activities and outcomes associated with the spending would strain many accountability frameworks.

Participating groups, member organizations, corporations, or public-sector organizations might fund such an institution entirely. Notwithstanding the inevitability of political concerns about staffing or geographic concerns, this issue would center on ensuring proper and consistent access methods that are impervious to most common forms of electronic attacks by both public- and private-sector organizations.

This is particularly relevant if the threat, risk or potential attack data are related to criminal offenses. The organization may be hampered by a restriction on the dissemination of information that is considered pertinent to an investigation. This may be the result of an ongoing investigation in which the information remains closely guarded by law enforcement for the sake of protection. It may also involve a requirement not to communicate using systems holding information in which the appropriateness of communications (including the integrity of the message communicated) is being challenged. Though the entire reason for these organizations is to distribute information throughout the enterprise in a collaborative effort to stop or prevent such attacks, there may be circumstances in which law enforcement activities will hamper and in some cases conflict with the net result activities of the member organization's primary charter.

In other words, such an attack or action may present the need for a lockdown as it would now be part of an investigation as evidentiary, thus restricting efforts to other interested parties within the ISAC that may not be directly involved with the investigation but may want or need the evidentiary information for preventive measures.

Another idea not originally considered is how the organization may be viewed as a liability in relation to the information it provides, disseminates, or distributes. This information may not be correct or may even be delayed such that further actions based on providing this information may lead to additional vulnerabilities if implemented.

That is, the bureaucracy of the participating member organizations or of any organizations that feed this information to the dissemination group or consortium may cause further vulnerabilities by the lack of timely dissemination of information that would have otherwise prevented such an attack from ever happening. Moreover, if the information is released too rapidly, it might not have been properly sanitized.[7]

11.11 Consequences of Accidental Disclosure of Information

It is possible to consider circumstances involving damages and liabilities caused by potential wrongdoings that might be misplaced, mismanaged, or provided through incorrect information or even through information that has been distributed in an untimely manner (either too soon or too late). This raises questions about the validity of the information in that it is communicated through an outside third-party organization (or even recognized as a participating member organization of the consortium or institution), another governing body, such as a government department, agency, or organizational entity, or even another corporation (not necessarily a participating member body organization).

Because of the complexities of these liabilities, their issues, or possible solutions, it may be difficult to say the method that should be used by the administrative process. To any potential issue relating to its method of dissemination or distribution of any information provided, the consortium or institution may need to comply with either government or corporate compliance or governance requirements. This is particularly true if any such information is not sanitized or may contain data that could be considered harmful or extremely detrimental to the reputation of the corporate entity, government agency, or department in question—or even worse, to its continuation and existence.

Without legal and operational assurances about preventing harmful legal or political embroilments or entanglements or provisions for the preservation of the data collected, the participating member organization might not be capable of fulfilling its organizational charter.

11.12 Intellectual Property and ISACs

With the introduction of intellectual property laws and standards within the past decade, ISACs may be hampered or constrained in their dissemination and distribution of information. Although it is unclear as to how these organizations may define a conclusion, it may be possible to generalize that a third-party organization providing attack information may forgo commercial rights to it.

Thus, the situation then becomes much like a feeding frenzy in that other interested participating member organizations or groups might find the relevant information not only useful but also valuable, thus making efforts or attempts to claim the property as their own. This is particularly true especially if the entity or participating member organization in question foresees any abilities in terms of revenue-generating capabilities, thus reinforcing the feeding frenzy scenario.

11.13 Trend Analysis

The key value to the repository of information involves the ability to conduct trend analysis. When sanitization has occurred, this will have to take two forms—a public and a controlled (or private) form. To the usual assortment of queries that can allow for a rather refined search, much of the value lies in the ability to identify issues at operational and strategic levels, both within single organizations and across communities. This is a matter of resource allocation, in which there are significant numbers of entities facing similar challenges. It behooves the overseeing organization to address those issues in such a way that it can reduce any levels of residual risk in the system, also demonstrating an appropriate allocation of resources.

11.14 Reporting Trends

Many systems designed to meet the aforementioned need are criticized in that much information goes in but no information comes back out to the

user. The goal is to have a system that evolves or can ideally begin to anticipate its environment. When the results (unclassified or nonnational interests)
are not communicated, a significant piece of the learning system is denied to
the end-user community.

Though reporting trends is important, another aspect of reporting
involves the reporting of progress. This will take place on two levels. As the
system becomes more mature, there should be a movement toward predicted
risk and observed risk in the system being state of equilibrium between each
other. Trend analysis focuses on the validation of the system's ability to identify and to calculate risk within the system. The level of risk in the system
should gradually reach a balance (infrastructure assurance versus capacity
management) and then should slowly move into progressively lower impacts
as the system eventually integrates resiliency and redundancy into itself, such
that single points of failure are ultimately engineered out of the system. The
latter point is particularly pertinent at operational and strategic levels when
single points of failure mean the loss of capacity across the full network.

11.15 Information Sharing and Definition and Categorization Challenges

Having looked at naturally disparate systems with divergent goals, it should
come as no surprise that there are two significant challenges when sharing
information. The first challenge involves communication of the actual message. Many systems do not share common definitions, and there has never
been a completely standardized approach to many of the issues. The result is
that confusion can arise when two systems, developed in two separate ways,
are required to interact and evaluate each other's data.

The second challenge involves the sheer mass of data and the ability to
sort them. A need exists to ensure a common methodology for identifying
the core characteristics of the communicated information that narrows initial searches but does not hide the data from more general searches. This
introduces the concept of data categorization as a part of a communication
tool. Such categorization can provide a key for area-wide communications
(at the distributed network and internet levels), searching for data containing
information that may be of value to the organization.

If we link this concept of data categorization to the internal communications of an organization, the trusted network, partners, and then the outside
world, we begin to identify the need for a system by which data might be
naturally tagged so that the data themselves become relatively smart. In this
concept, data are able to sense their locations within the network topography
and as well as the other data-based activities around them, ensuring that
they maintain their core characteristics (i.e., confidentiality, integrity, and

availability) along their intended design. This would not be possible without a clear sense of the trusted network, the agreements in place, and the value of the information across the system.

11.16 ABC Transport

ABC Transport has, within its corporate policy, requirements that learning systems are included across its organization, particularly with respect to CIP. For each of the following, an investigation or inquiry must take place and the results communicated back to management so that threat and risk assessment may be validated as having taken into account the information provided:

- Injuries to personnel
- Loss of access to facilities, including internal spaces, due to influences that are internal (e.g., strikes) or external (e.g., natural disasters)
- Loss in terms of the payments to restore services or material due to destruction, damage, or disruption
- Cyber events, such as viruses or malicious software detected on the system
- Losses in productivity due to failures in anticipated services
- Failures in various plans, including the CIP plan, due to incidents or events through the observation of drills, exercises, or routine operations by either facility personnel or clients

This cycle parallels processes formerly used for total quality management systems that involved all aspects of the organization. Each employee is required to report a deficiency but also to include a suggestion or recommendation as to how to alleviate the issue. The company has also included an award, presented at the end of the fiscal year, for the individual whose suggestion resulted in the best returns for the organization.

This information goes to the regional office, which then determines whether there are common failings in the system or areas within the system that appear to be having greater difficulty. These are prioritized first in terms of investigation and then of capacity building to certify that baseline conditions are actually maintained and not simply reported.

At the operational level, office managers meet monthly to discuss challenges and possible solutions. This is generally handled through teleconference calls but can also occur face to face depending on the topic, its sensitivity, and the need to investigate further.

Similarly, the operational-level management participates with the organization's regional security and other committees to better understand the environment—before attending speaking points are identified and signed off by the regional manager or point of contact at the strategic level.

At the strategic level, regional and office managers have access to the repository of information, which performs the following functions:

- Generation of lists (in priority) indicating where the system has demonstrated that it will likely fail; the intent of this listing is to generate additional lists of topics that need to be addressed
- Solicit advice from other managers at the same level to propose either alternatives or alternative directions for addressing complex problems in such a way that the system does not simply provide "cookie cutter"[8] answers applicable in another area
- Provide trends to senior management showing which areas are more likely to be at risk. Senior management presents two awards. The first is to the regional team that is best able to reduce risk in the system. The second involves the individual who contributes innovative solutions to challenges. Senior management presents these awards to the individuals or teams involved, and personal rewards are included in each.

11.17 Questions

1. Do you feel that there are better methods by which information could be shared unilaterally among all interested parties without legal situations arising? Explain.
2. Should the government take more or less active role in playing traffic cop with the information-sharing organizations? Explain why and how.
3. Explain the relationship between the internal organization, the trusted network, and outside competitors in relation to the sharing of information. Consider failures in the system and how those would relate to a vulnerability to persist within the system.
4. Do you feel that the lessons learned program has been a success or not? Explain why and how.
5. Identify three databases of information provided to a central repository, processed, and then released. Describe the community to which the information is released and how the information presented differs from the information released. Is the value of the information retained?
6. Describe the value of categorization and information sharing in this kind of environment in terms of the ability to prevent, detect, respond to, and recover from events. Does this have an impact on the overall stability of the system? Explain how.

Notes

1. Refer to Chapter 5 in this volume for more information about the C-I-A triad.
2. ISACs were voluntarily created to provide an information sharing and analysis capability to support their members' efforts to mitigate risk and to effectively respond to adverse events, including cyber, physical, and natural events. ISACs have been established within most of the critical infrastructure sectors identified in federal policy, including those for banking and finance, chemicals and hazardous materials, drinking water and water-treatment systems, emergency services, energy, food, government, information technology and telecommunications, and transportation. ISACs have also been established for other industry sectors, including real estate and research and education networking. U.S. General Accounting Office, "Critical Infrastructure Protection: Improving Information Sharing with Infrastructure Sectors (GAO-04-780), July 2004, http://www.gao.gov/new.items/d04780.pdf.
3. Paul Roberts, "9-11 Commissioner Calls for End to ISACs," *InfoWorld*, February 18, 2005, http://www.infoworld.com/article/05/02/18/HNsecurity911_1.html? DISASTER%20RECOVERY.
4. Committee on Homeland Security, "Information Sharing," *The State of Homeland Security 2006: An Annual Report Card on the Department of Homeland Security*, p. 36 (on-line page 46), http://homeland.house.gov/SiteDocuments/20060814122421-06109.pdf.
5. ibid.
6. The term *capacity* is not to be confused with capacity management of a transportation system but is instead associated with capacity of participation of an entire corporation or organizational entity participating in a joint or participatory consortium to exchange and share ideas.
7. The term *sanitized* refers to "data sanitization" in which critical, or critical key, information is removed for anonymity reasons; thus, any individuals or organization that could potentially be at risk would not be exposed, especially those working with public forums or in areas of public display or disclosure.
8. This replication process utilizes a "master cookie," which is replicated to other locations, either geographically or functionally, in an effort to streamline greater efficiencies, to achieve improved security, and to improve monitoring capabilities by using the replicated process, procedure, or method.

Fragility and Fragility Analysis Management 12

Objective: At the end of this chapter, the reader will understand the following:

- The concept of *fragility* and how it can be used to determine business failure
- Method by which you may be able to ascertain fragility within your organization through measurable results
- Data categorization concepts and how data become information and are increasingly more important and valuable to the enterprise organization

12.1 Introduction

The ability to evolve and to predict necessary evolutionary steps ultimately dictates whether the system flourishes or begins to decay. The difference between the rate at which the environment changes and the organization changes will have a significant influence on how much effort the organization must make to flourish or even survive. The four key factors are resiliency, redundancy, robustness, and fragility.

Three of these factors describe the ability of the organization to withstand the shock of impact, whereas fragility describes vulnerability of the system against such a shock. *Resiliency* is used in a sense that mirrors business continuity planning (BCP) in that it describes the ability of something to "take a hit," and then to reestablish acceptable levels of service. For those with a martial arts background, this is very much the difference between the willow and the oak. The oak snaps under the pressure whereas the willow bends under it, springing back to its original form. *Redundancy*, on the other hand, describes a condition whereby the organization is able to establish options to replace the service. This means the elimination of single points of failure (SPoF) through allocating internal resources or entering into appropriate agreements with outside organizations so that the organization simply switches to a new source of the same input. *Robustness* differs slightly in that it is relatively passive compared with the first two. This concept describes a condition in which the design and system implementation can overcome attack considered through sheer strength. An attack cannot overcome design characteristics or safeguards in place, much like approaches intended by castle walls. *Fragility*, though, describes a condition in which the organization does not possess these to an adequate level and may even lack the ability to detect or respond to changes in the environment that could indicate that such an attack was imminent.

Understanding fragility lies with the notion that CIP for the transportation system does not lie primarily with the protection of a given infrastructure but with the maintenance and protection of capacity across the transportation system. The value of the infrastructure lies in what capacity it contributes or supports the overall system. The asset value of the infrastructure provides a secondary role in the nature of loss, which might have far-reaching implications about the infrastructure's ability to maintain capacity in the long term.

12.2 Requirement for Information

Managing the capacity of the transportation system at any level requires an understanding of levels of capacity present within the system. This has three aspects. The first aspect deals with how the system is performing and to what degree capacity surpluses or deficits exist within the system. The second aspect deals with the same consideration for inputs necessary in maintaining that capacity, both from a supply chain management and interdependency point of view. Decisions about how to manage the transportation system may be made on a reasonably firm foundation when the information is relatively up to date and has a high level of integrity (i.e., the information is accepted and trusted). When this information becomes dated or untrustworthy, risks associated with inappropriate or incomplete strategies may increase significantly. The third aspect deals with the ability to identify and seek out repositories of trusted information in response to unforeseen challenges or situations. It is reasonable to assume that every organization is able to generate information repositories and to keep them up to date. The most significant challenge is how the system can identify, collect, collate, process, and disseminate this information to its best use.

12.3 Repositories of Information

The network of trust and communities of interest become very valuable since they can identify repositories of trustworthy information. This requires a level of balancing supported through memorandums of understanding (MOA) or mutual aid agreements. On one hand, there are challenges in terms of resiliency, robustness, and redundancy as these factors reside outside of the organization's main sphere of influence. On the other hand, gathering reliable information from those trusted sources allows for an efficient input supporting risk management and similar activities. Repositories may be compared with each other to identify potential service delivery gaps. Consider, for example, a database maintained by an emergency call center that tracked how long it took emergency vehicles to get to a scene. This could be compared with the security logs indicating that emergency vehicles have arrived at the scene. This can be cross-referenced to determine the integrity of the data through a simple validation process.

These repositories of information reside naturally within the organization. The key is understanding where repositories lie, who has control over their use, the restrictions on their use, and the meaning of the data contained within them. Loss prevention programs, maintenance programs, and other similar kinds of activities will all record this information for their own purposes. Loss of integrity is a risk stemming from differences in definitions and approaches resulting in errors similar to translation errors between languages. Thus, care must be taken when combining systems to understand and to document the definition bases and doctrinal base so that the meaning of the data is understood and can be reliably processed into information and then intelligence. Once again, this reveals the need for coordination and cooperation between the various components within the processes and system.

An organization would be well advised to keep the following kinds of data in its repositories of information:

- Performance metrics
- Maintenance records
- Records associated with closures due to inclement conditions
- Records associated with disruptions due to, for example, spills or accidents
- Security events
- Safety events
- Loss prevention and purchasing to replace, repair, or restore services
- Any other costs associated with meeting obligations arising from abnormal operating conditions
- MOU or mutual aid agreements

One factor of fragility deals with the ability (or inability) to identify repositories of information. These generally indicate contributions of personnel, objects, facilities, organizational information, or activities within the system in terms of their performance, capacity, disruptions, and loss. For these structures to be effective, they must be applied consistently using a measurement system that can be defined, documented, and repeated as required.

This chapter examines several aspects of fragility. These aspects pertaining to each other consist of persons, assets, facilities, information, and activities and how they contribute to successful process completion. Where they are identified, a reasonably complete picture may be compiled even though fragility may still exist within the system. As one or more of these may not identified or lost to the process, an understanding becomes less than complete, allowing for an increase in fragility. The first element determines whether or not the organization has taken into account all of these as they work toward understanding of fragility in their environment (Table 12.1).

Consider a situation in which ABC Transport must determine when to have certain parcels available for distribution. On one hand, the distribution center could set a schedule and then conduct its own operations in accordance with that schedule. This leads to a situation in which one part of the system can demonstrate that it is meeting its accountabilities (maintaining the schedule) but does not include various factors associated with the personnel, assets, facilities, information, or assets actually performing many of the functions demanded by the distribution center. This kind of approach reflects a situation in which several of the elements previously outlined are not taken into consideration and are lacking in the decision-making processes. Thus, this approach illustrates an approach replete with fragility.

ABC Transport might choose to ensure that it is able to identify the status and capacity of the personnel, assets, facilities, information, and activities it relies on to meet its objectives. This may involve physical, technical, procedural, or other measures identifying changes in performance and communicating those changes at the earliest opportunity so that risks associated with disruptions are minimized. For example, the schedule of any given system might link to the driver's electronic devices to indicate where he or she is on his or her route. Where the driver deviates from the route's intended design, the device could communicate back to the distribution office that the driver

Table 12.1 Elements Considered

Weight	Description of Weighed Element (Persons, Objects, Facilities, Information, Activities)
5	No element of the five elements is present; all five elements are missing.
4	One element is present; the other four elements remain missing.
3	Two elements are present; the other three elements remain missing.
2	Three elements are present; the other two elements remain missing.
1	Four or more elements are present.

is behind or ahead of schedule, allowing for a reduction in the disruption of activities by providing more advance notice to other processes that might rely on the presence of the driver or his or her deliveries. This system, compared with the former and unlinked systems, has greater resiliency and much more robust processes.

12.4 Lines of Communication

The value in sharing repositories lies in data generated by a trusted source and maintained due to a vested interest. Each party has an interest in the accuracy and completeness of the data because they provide an input into each party's processes. In the case of a shared repository supported through mutual agreements, the holder of the data may share them with the understanding that they will be handled a certain way. Though intelligence information that a certain individual may be entering the country on a plane or ship is valuable, if it arrives three weeks after the arrival of an aircraft or ship in question the value of the data is considered limited. In essence, there are two key areas of risk within the system:

1. Loss of integrity: Data cannot be trusted by those who are assessing them; rendering is less than useful.
2. Loss of availability: Data are not present when they are needed.

The value of these two considerations will likely change depending on the specific nature of the information and where it needs to be put into the process. Looking at these two factors, the next scalar for the rating of fragility within the system might be generated as shown in Table 12.2.

Returning to our delivery system, the weight would depend on the design of the system. A strong system might provide updates at specified intervals as

Table 12.2   Data and Information

Weight	Description of Weighed Element
5	No availability; acceptance of data is not possible.
4	Data are available, but acceptance of data is delayed beyond usefulness of the data.
3	Data are available, but ability to assess data is possible—with significant delays.
2	Data are available, and ability to assess data is possible but with some effort.
1	Data are available, and acceptance of data is intrinsic to the communication process.

well as whenever a change occurred. A weaker system might send periodic or even infrequent updates (e.g., twice a day) and have no automatic way for the driver to accept the message, resulting in the dispatcher's having to contact the vehicle to work out the scheduling information.

12.5 Data Categorization

As one progresses from the tactical to the operational and then finally to the strategic level, the amount of data, information, and intelligence being processed increases significantly. It is possible that this could lead to a situation in which the organization cannot maintain the ability to use the input to its best effect—when either valuable information is missed or the data imported are not completely understood. Categorization of data, therefore, becomes increasingly important, particularly when vital information is being communicated, to assure that it receives the attention it deserves.

Categorization hinges on the acceptance of common criteria across a community or to the ability to relate to common criteria (often used by industry). This is relatively simple when working within a single administration, as it is simply a matter of the appropriate accountable individual making the decisions as to categories. It may even be reasonably manageable within a trusted network in which membership is contingent on setting up and using systems addressing this issue. The issue becomes very complex when dealing with organizations that are able to operate in isolation and in which no benefit or sanction exists because of the isolation. As the need for compromise among these parties increases, so does the complexity associated with establishing, implementing, monitoring, and maintaining that common criteria.

Tools used to sort the data also affect the ability to benefit from categorization. Mechanical or automated systems may not benefit from the necessary level of human judgment whereas human-driven systems may be subject to failures or errors in judgment. This risk is unavoidable within the system but may be mitigated through procedural controls, such as the periodic validation of results or the spot-checking of results. The keys here are to certify that the data are categorized correctly and then sorted using the best option.

Fragility, therefore, can be described in terms of the inability of the system to effectively manage data being presented or communicated either within itself or with its trusted partners, which can be described or prioritized using Table 12.3.

Returning to our delivery driver, the data may be categorized as *delayed*, *delayed due to traffic*, or *delayed due to mechanical failure*. Categorizing the

Table 12.3   Compatibility of Data Categories

Weight	Description of Weighed Element
5	No categorization; data volume overwhelms the system.
4	Categorization criteria are incompatible across multiple layers in the system.
3	Categorization criteria do not align directly, resulting in some limited confusion.
2	Categorization criteria are compatible and relatable with some effort.
1	Categorization criteria become integral and are consistent across the entire system.

data in this way changes their value. At the same time, a significant value to the organization is provided when data are transmitted in such a way that systems tracking these issues can detect their presence, identify their values, and validate their contents.

12.6 Adaptability of the Categorization Process

Appropriately, categorizing data requires that certain predetermined policies, standards, guidelines, and procedures are followed so that results are both consistent and comparable. Categorizing data can also fall prey to conditions in which the system does not easily adapt to changing environments. Given the current fluid environment, the need to develop and include flexible systems that can respond appropriately to unforeseen events is becoming increasingly important. Rigid systems tend not to address unpredictable events, instead focusing on a snapshot in time. This can result in intentional errors made as users attempt to find the best fit for the data entered. As the number of changes missed increases, the integrity of the data becomes increasingly questionable.

Allowing change into the system can lead to losses in control that threaten the integrity of the overall system. Carefully controlling and delegating the change process becomes ever more important as the ability to enter data or make changes spreads throughout the system. As this ability expands, errors (intentional or otherwise) can be entered into the system.

Fragility, in this respect, reflects the inability of the system to change in a controlled fashion. This inability means that the integrity of the categorization process is lost, either through mislabeling or through the failure to detect past instances that need to be adjusted into the new category. This could be reflected in the scalar system shown in Table 12.4.

In our example, the central distribution center has a system to track why drivers are late. This includes a number of set categories—persons not expecting parcels, objects missing or damaged delays at clients' locations, missing or

Table 12.4   Adaptability of the System

Weight	Description of Weighed Element
5	Adaptability is neither considered nor integrated into design or application of the system.
4	Adaptability is not considered, and ad-hoc methods are used to adapt the system.
3	Adaptability is considered but is entirely formalized within the system.
2	Adaptability is considered and documented, but policies are not enforced throughout the system.
1	Adaptability is considered and fully integrated into the system.

erroneous information, and other activities that do not lead to the intended results. Within each category, a number of incidents is possible. Although categories may be selected from a drag-down list, drivers simply enter a text message into the pad. The system has been configured to search for the applicable data and sort it accordingly. A more fragile system might involve a system that does not have the same level of centralized (and therefore consistent) controls present. This deficiency results in conditions in which parts of the system begin to operate independently.

12.7 Adaptability of Data Sets or Mutability

Categorization enters a new risk into the system. Data may not be categorized either appropriately or completely. Thus, another key to maintaining the value of data collected over time will be the ability to go back into the system in which data reside to reapply rules or policies that recategorize them. It would also be necessary to have the confidence that those rules had been applied completely.

Applying these rules can pose a relatively complicated logistical challenge within the system. Imposing the system from a central trusted source exposes the system to a strategic-level risk in that attackers spoofing[1] or piggybacking[2] on trusted signals would gain a conduit across the overall system, resulting in much higher damage than if the system had a level of compartmentalization. Another approach would require the creation of a standard, such as those used for information management and information technology (IM/IT) security, to describe systems used within the transportation system so that such systems were assured in terms of their ability to relate back to central trusted sources in which assessments can take place. The key challenge in this respect lies in the legal, economic, and operational impacts associated with the compulsion to deploy systems within the free-market environment but without creating the appearance of undue government interference.

For the distribution center, this means that a new categorization includes a keyword search. Where these keywords appear, it must be determined if the record should be included in the new category.

12.8 Assessment

Data have little value in and of themselves. It is through their assessment that data provide value to the system (which we refer to as *information* and eventually *intelligence*). The first phase of assessment is essentially the collation of data for purposes of evaluation, driven by the categorization process. The second phase involves identifying changes in capacity (represented by operating levels), criticality, threat, impact, risk, vulnerabilities, and safeguards within the system. The third phase entails determining importance of these changes to the organization using a trusted analytic process by appropriately trained and experienced individuals. Feeding this information back against the threat and risk assessment and then validating the results provide a significant step toward maintaining the ability to evolve.

The assessment process may be subject to a number of factors. The first is the availability of capable personnel that collate information and assess it, whether hands on or through the monitoring of electronic processes. The second involves the timeliness of the results of assessment process to the decision-making process, a factor that will figure heavily with the discussion to follow shortly regarding how disruptions and failures can cascade throughout the system. The final challenge encompasses the ability to transform raw data into what may be considered intelligence within the system and to identify the usefulness of that intelligence to the organization's best advantage.

The level of fragility within the assessment process is a factor of quality and time. Systems that can produce high-quality assessments of significance that are based on appropriate levels of data could be considered reasonably strong. Systems that fail to deliver this assessment or that have significant degradations in the quality of assessments because time constraints allow for less analysis increase the levels of fragility within themselves. This could be represented within the scalar system shown in Table 12.5.

Validation of processes in this context has a two-step process. The first involves determining whether the full assessment process had actually been followed, including a verification of data being used as being trusted and complete. Having determined that data were assessed, the next step guarantees that the assessment process was consistently applied every time. This function would likely have to be undertaken by an outside organization that could impartially review the decision and investigate the impact.

Table 12.5   Quality Assurance

Weight	Description of Weighed Element
5	Processes are not validated and applied by individuals of uncertain capability under time and resource constraints, preventing quality assurance of intelligence.
4	Processes are not validated and applied by individuals of limited capability under time and resource constraints, minimizing quality assurance activities.
3	Processes are not validated but are applied by capable individuals who operate under time and resource constraints, allowing limited quality assurance activities.
2	Processes are validated and applied by capable individuals who operate under time and resource constraints allowing for routine quality assurance activities.
1	Processes validated and applied by capable individuals who operate with time and resource constraints that do not affect quality assurance activities.

12.9 Integration into Mitigation Strategies

Intelligence provides an opportunity for the transportation system to identify conditions in which new mitigation strategies or approaches may be appropriate. Accomplishing this means that the organization has to move beyond the ability to collect information (as previously described) and into the realm of processing that information to determine its true significance to the organization—and, in many cases, out into the transportation system itself. The goal is to develop the ability to integrate this intelligence into the system. This involves the ability to communicate that intelligence back to the right parts of the system (or across the system) appropriately so that those who allocate resources may take necessary action to put those strategies or approaches into practice.

Communicating intelligence back into the transportation system becomes the first challenge. What form should the communication take? This decision will be driven by three major factors:

1. Scope: How much of the system is involved or can be communicated with?
2. Time: When does the approach have to be in place?
3. Resources: If there are costs involved, how will they be paid?

A fourth factor operates in parallel to the three factors just outlined—the sensitivity of the information. When such information is generated through classified or other means, unauthorized disclosure of this information could have serious ramifications. As a result, its release cannot be legitimately approved except under very tightly controlled and balanced conditions involving the author or authority responsible for the information. This in itself poses a challenge when looking at facts discussed earlier. Much of the infrastructure is owned by private sectors with significant involvement by stakeholders (e.g., local, municipal, provincial, territorial, state, national, international) that may not be adequately integrated into systems allowing for passage of that information.

Assuming that the assessment process is of reasonable quality, the next factor in fragility within the system becomes the barriers preventing that assessment from reaching the appropriate decision-making process or processes (Table 12.6).

It should be noted that external controls are considered more difficult to influence, as accountability and risks associated with maintaining those controls are assumed by external organizations. Relaxing those controls will require some level of argument that it is appropriate to do so. Internally, management has significantly more control over, for example, its own internal policies, which allows those decisions to be made given the balancing of the various risk appetites involved.

Having received intelligence, the next step involves using that intelligence in decision-making processes and then in the integration of the decision, based on the assessment, into the transportation system. This creates two potential issues: (1) the fact that not all decisions will be universally applicable; and (2) the ability for the decision to be amended should it be discovered that it is not having the desired outcome. This will be dependent on perceived risk, organizational structure (including delegations), and corporate culture within the organization making the decision. Some organizations may have entrenched cultures and are resistant to change, whereas others may embrace change as part of doing business. As a result, it is likely that a variety of different methods will be apparent across the system. Some will involve compulsion (e.g., through regulation) whereas at the other extreme some might

Table 12.6   Ability to Communicate Between Partners

Weight	Description of Weighed Element
5	Barriers to communication exist externally, and the organization has no ability to influence those barriers.
4	Barriers to communication exist externally, and the organization has limited ability to influence those barriers.
3	Barriers to communication exist internally, but the organization must expend significant efforts to affect barriers.
2	Barriers to communication exist internally, and those communicating have the ability to seek direct approval to bypass or affect those barriers.
1	Barriers to communication exist internally, but those communicating have the ability to manipulate those barriers appropriately.

be possible through voluntary efforts because the benefits are apparent to the participants. When looking at a management cycle that pertains to analyzing, planning, organizing, and controlling resources, the following four considerations will factor heavily when attempting to assess the potential for accomplishing the necessary change:

1. The level of compulsion is appropriate to the demand for integration.
2. Those integrating the changes are capable and delegated to do so.
3. Mitigation strategies or approaches are translated into the appropriate useful and measurable format (e.g., standards, guidelines, procedures).
4. The useful information is communicated into the trusted network and becomes part of the conditions to remain within the trusted network.

Fragility, in this respect, indicates conditions in which the system acts in opposition to those four criteria in such a way that its integration either causes unnecessary harm within the system or cannot be verified as being in force within the system. This may be reflected as shown in Table 12.7.

Table 12.7   Mechanism Needed to Compel Participation

Weight	Description of Weighed Element
5	The choice of mechanism is based on administration and not results. The decision is imposed on communities that will remain part of the trusted network even with the new risk being added to the system (no validation of changes made).
4	The choice of mechanism is based on a balance of administration and goals with the decision imposed on communities that receive limited support for a finite period to maintain their status within the trusted network (validation of changes not necessarily required).
3	The choice of is mechanism based on goals with the decision imposed on communities that receive limited but stable support to maintain their status within the trusted network (validation required).
2	The choice of mechanism operates within an administration that recognizes goals as the primary objective of its activities and that provides support for the new requirements as part of its obligation to the trusted network but with conditions to be met.
1	The administration of the decision is goal driven but follows validated checks and balances. Decisions are communicated to communities receiving support for the measures as part of the requirement to comply with the decision and that are integrated into long-term plans as a condition of both membership in the trusted network and maintaining the authority over the system.

12.10 Addressing Capacity in Decision-Making Gaps

The transportation system has no clear or single owner, leading to a situation in which no particular organization can legitimately speak for the entire system. Similarly, consequences associated with any impact within the system may have an even more significant level of impact on society, compelling government intervention (if only from the perception of perceived risk). Because this authority is often unclear, decisions made (both in terms of the measures to be taken and in terms of the mechanism used to implement them) may be subject to one or more challenges from a variety of sources.

The inability to identify these gaps, to chart courses of action, and therefore to avoid challenges against the legitimacy of the system provides another level of fragility in the system. In this case, fragility resides in the fact that individuals, with a variety of goals (some of which may be the removal of safeguards), can challenge the frameworks that are in place to assure infrastructure and capacity through legal and other means. This attacks the stability of the system's framework, particularly of any safeguards and might be described as shown in Table 12.8.

12.11 Translating of Strategies into Action

The translation of a mitigation strategy into specific procedures required to mitigate risk involves identifying how to measure the success of the mitigation strategy. This involves translating theoretical goals into quantifiable, measurable, and repeatable results. It also involves validating this process to track changes associated with specific actions taken.

Table 12.8 Ability to Implement Evolutionary Steps

Weight	Description of Weighed Element
5	Authoritative base for the framework cannot identify requirements and is subject to broad challenges of its authority.
4	Authoritative base for the framework can identify requirements but faces significant challenges developing and implementing courses of action.
3	Authoritative base for the framework can identify requirements and develop courses of action but may lack the political will to implement them completely.
2	Authoritative base for the framework can identify requirements and develop courses of action but may take significant time in implementing them.
1	Authoritative base for the framework can identify, develop, and implement courses of action without significant challenges to the framework.

The translation of strategy or approach into action requires that individuals or organizations within the community who are required to impose changes within the system are in fact capable of performing such acts. This is very similar to the requirement for those transforming assessments and recommendations into mitigation strategies, but on a broader level. The prior deals with those that are setting a broader direction to address, transfer, accept, or ignore risk within the system. The latter identifies specific requirements or procedures to be developed, implemented, monitored, and adjusted to achieve the goals of the mitigation strategy.

Fragility, in this respect, is a condition involving the lack of necessary expertise within the system to accomplish those goals. This expertise becomes the enabler that links the conditions identified through the strategy and translates them into specific procedures, making certain that the conditions identified at the start of the process are monitored for change. The second level of expertise is the ability to understand and apply more than simply dogmatic approaches when looking at conditions that could influence what specific steps need to be implemented. These conditions can be demonstrated as shown in Table 12.9.

Similarly, one must be careful in using the term *expertise*. At one level, formalized programs from trusted institutions subject to outside review may have fully credible programs that offer significant value. At the other end of the spectrum, individuals with little or no experience that have identified a market may also be practicing within immature industries and taking advantage of the lack of codified knowledge against which they would be assessed.

12.12 The Rough Fragility Score for Evolution

The conditions in Table 12.9 comment on the ability of the organization to evolve based on the discovery of new information within the system. This

Table 12.9 Maturity of Expertise Relevant to Industry

Weight	Description of Weighed Element
5	Industry is immature with no codified practices, and expertise is not present.
4	Industry is maturing, but codified practices do not exist and expertise not yet present.
3	Industry is establishing itself with limited codified practices, and a broad base of emerging expertise is starting to exist.
2	Industry is establishing itself with maturing codified practices, and a broad base of stable expertise exists.
1	Industry is established with codified practices, and a broad base of expertise exists.

ability within the CIP domain is of critical importance looking at the overall effectiveness of the system at any of the management levels.

The results from the rough scalars provided herein provide a hasty ranking that identifies, using a consistent methodology, levels of fragility with respect to the system's ability to evolve in response to new conditions (Table 12.10).

Using the ABC Transport example once more, ABC Transport tracks any divestitures associated with personnel, objects, facilities, information, or activities as part of a loss prevention program that calculates costs (impact) associated with any losses. This information was linked to a centralized BCP and continuity of operations (COOP) process responsible for the conduct of threat and risk assessments. Given that the system is just evolving, it is determined that data would be categorized loosely for the time being but would be watched to guarantee flexibility within the system. This information, gathered in the assessment process staffed by certified professional members of an accredited body, are responsible for proposing the mitigation strategies to management. Management signs off on these recommendations following a conference or teleconference with the regional managers to verify that no capacity or capability issues exist in the operational-level context. Finally, each regional manager has access to a number of experts to determine available options and to assist in the integration of the system.

This is based on the following application of the scores:

$$(A + B) \times (C + D) \times (E + F + G) \times H \times J$$

$$= (1 + 1) \times (3 + 2) \times (4 + 4 + 2) \times 2 \times 4$$

$$= (2) \times (5) \times (10) \times (8)$$

$$= 800$$

If there are no lines of communication between these systems (i.e., information is not being communicated), the following shift would result:

$$(A + B) \times (C + D) \times (E + F + G) \times H \times J$$

$$= (1 + 5) \times (3 + 2) \times (4 + 4 + 2) \times 2 \times 4$$

$$= (6) \times (5) \times (10) \times (8)$$

$$= 2400$$

Table 12.10 Levels of Fragility

System	Repositories	Communication Lines	Data Categorization	Adjustment of Categories	Assessment	Intelligence to Mitigation	Mechanics of Decision Making	Capacity Gaps	Strategy and Approach to Action	Score
Figure	2.1	12.2	12.3	12.4	12.5	12.6	12.7	12.8	12.9	
Label	A	B	C	D	E	F	G	H	J	
Name	1–5	1–5	1–5	1–5	2(1–5)	2(1–5)	1–5	1–5	2(1–5)	
Example 1	1	1	3	2	4	4	2	2	4	800
Example 2	1	5	3	2	4	4	2	2	4	2,400

This method of scoring is one way of determining fragility in a system, with higher numbers indicating increased fragility. From Table 12.9, the highest scores represent conditions that are most conducive to the appearance of fragility within the overall system. The aforementioned formula is described in Table 12.10.

12.13 Additional Factors with Respect to Fragility

Fragility is affected by four additional factors that operate beyond its ability to react to changes within the environment.

The first factor is geographic in nature. Given that the demand for movement must pass through nodes and conduits, the physical environment plays a significant factor by limiting the options associated with their creation. At a strategic level, this may involve the fact that an area restricts the creation to a single conduit. Features along that conduit (e.g., bridges, tunnels) can also act as single points of failure if something happens to cut that conduit. The desired outcome is for a situation in which, if the main connection fails, adequate capacity across other alternative connections is found to allow for continued movement. The lack of such routes or the inability to use these routes become an indicator of fragility of the system.

The second factor to affect fragility is the determination of whether the transportation system is reliant on outside factors to recover or to maintain its resiliency. This question revolves around issues similar to sovereignty over a critical infrastructure. If, for example, the transportation system maintains the necessary personnel, assets, facilities, information, and activities within itself, it can allocate those resources as needed or even on demand to maintain its resiliency. This control is lost as one moves outside the organization reviewed. For example, if those inputs are controlled by a foreign power it may achieve a level of economic or political advantage through the threat of denial of availability of those resources unless certain favorable conditions are met or concessions have been made. Of course, if these discussions are happening during a period of significant disruption, there will be significant importance attached to those discussions to reduce any impact. The presence of this condition represents another layer of fragility within the system.

The third factor relies on the disposition of interdependencies throughout the system. This operates at two levels. First is the general location of the origination of the interdependencies in support of the transportation system. Where they originate from outside of various common levels of administration, spans of control may be lost. Where they originate all from one location, the system may become vulnerable to a significant disruption within that area, due perhaps to a major natural disaster. The second level involves deter-

Table 12.11   Maintenance of Structure Integrity/Norms

Weight	Description of Weighed Element
5	No coherent maintenance program or willingness or ability to share costs
4	Incoherent maintenance program with limited willingness to share costs
3	Maintenance program established but subject to cost-sharing negotiations
2	Maintenance program established and cost sharing generally present but not integrated into long-term plans
1	Maintenance program established and cost-sharing process normalized and integrated into long-term plans

mining where single points of failure exist between the interdependencies and the overall system. This will affect all management levels and relates not to geographic locations but to how those interdependencies affect abilities in maintaining capacities and resiliency at each of those levels. Fragility exists within the system in which these interdependencies act to reduce inputs across the system or in which the disruption of the interdependency actually feeds back upon itself as the transportation system might provide a necessary service to another sector.

Maintenance of the infrastructure is the fourth factor that plays a crucial role in determining fragility of the system. The failure to maintain the transportation system, particularly looking at bridges and similar structures, leads to a level of unpredictability within the system as the infrastructure does not behave or survive as anticipated. Rebuilding the whole infrastructure today would not be feasible due to the sheer volume of material required as well as costs associated with the work. Thus, a reasonably well-coordinated effort across all levels of government (which share ownership across much of the infrastructure) and industry is required to maintain expected levels of capacity within the system without running the risk of an unforeseen failure. This aspect of fragility also poses a hidden risk when looking for excess capacity in the system as the planners or system could unintentionally overburden the system, leading to another failure (Table 12.11).

12.14 Rating Geographic, Sphere of Control, and Interdependency Fragility

The four factors reviewed in Section 12.12 play a significant role in determining levels of fragility within the system and can operate at any of the tactical, operational, or strategic levels. What needs to be clear is the ability of these factors to influence the ability to restore the capacity within the transportation system, not simply replace a piece of infrastructure (although that may be a key component in restoring the capacity).

Similarly, the protection of resources and the need to protect those resources enters a level of fragility into the system. This is because even though information has been received, the ability to respond to that new information is simply not there. This is particularly true in systems that trend toward constant outsourcing for alternatives and do not maintain their own level of internal capacity.

A parallel condition exists in the just-in-time supply chain system. Even though the information regarding a disruption is received, lack of adequate stocks of inputs means that information is often of little value. In this situation, the concept of fragility also has value when one factors in the reliability of the suppliers sending the input material. Where fragility factors score highly, the supply chain manager may want to recommend the maintenance of stores of inputs that are adequate to cover foreseeable disruptions. These stores would be replenished once the delayed shipment had arrived. This just-in-case supply chain would be an operational safeguard against disruptions due to loss of inputs.

The goal is to avoid networks that attempt to use linear tools to establish stability. For example, bulk sale discounts are often used to make sure that the best financial value may be achieved during the purchasing process. This involves committing all purchasing activities to one source. This linear approach creates a single point of failure within the interdependency model. It also breeds conditions necessary for the creation of what Holling[3] terms a "poverty trap" in the adaptive cycle of a system through the erosion of interconnectedness with other parts of the system's other interdependencies that then further lock that lack of connectivity back into the system as part of the contracting process. Similarly, where monopolies occur, a similar condition might exist in which the management level of the monopoly may begin to exert influence into the system through the threat of disruption or the demand for less connectivity to maintain trusted supply lines. The key is to assure that there are the necessary levels of connectivity across all interdependencies so that the disruption of any particular interdependency cannot have a catastrophic affect on the system's ability to deliver its service.

Although complex mathematical tools may be available to map the influence of these factors, one has to realize that to be deployed, such tools would have to operate at the back end of systems or within relatively cloistered groups of experts. A simpler rating system needs to be devised to map levels of fragility within the system in such a way that the average person operating a business or planning an operation can actually determine the levels of fragility. One such option might be the scenario shown in Table 12.12.

Table 12.12   Fragility of Links Between Physical Networks

Weight	Description of Weighed Element
5	Movement between networks on single main routes that have multiple SPoF and conditions exist that could cause disruption (natural events such as landslides into tunnels).
4	Movement between networks on single main routes that have SPoF but for which preventive measures or response measures are in place to minimize the disruption caused by the failure in the system.
3	Movement between networks has options to use one of a number of main routes, but routes must be chosen well in advance to avoid SPoF or disruptions.
2	Movement between networks has options to use a number of main routes, and information is routinely communicated regarding the performance of those routes and the conditions that could cause a disruption. These routes offer flexibility in choice for routing.
1	Movement between networks has multiple, viable options that are clearly understood and that have the possibility for shifting routes quickly upon discovery of the conditions that could lead to disruption.

For example, in examining at a major east–west route of a country that went through a mountain pass and was the sole route available, it might be argued that the factor involved at this point would be between 4 and 5 and would lean toward 5 when the natural conditions that caused landslides or similar events were present. If two such routes were available that were not subject to the similar or same conditions, that factor might reduce to a 3.

With respect to interdependencies, a slightly more complicated approach needs to be looked at when rating these factors. The challenge here involves balancing the need to reflect interdependencies in terms of its being a single point of failure, and that interdependencies might cycle back into themselves from their own reliance on the transportation system—perhaps increasing the potential rate of disruption.

In this case, a factor similar to $2r{:}n$ may be more appropriate, where r represents the reliance on the sector looked at within the transportation sector to deliver its own services and n represents the number of independently operating service providers capable of providing a service to the transportation sector. This might use the qualitative rating scale shown in Table 12.13.

Under this system, if two service providers ($n = 2$) were able to meet the needs of the interdependency within the area but if both relied on a part of the transportation system that would be directly affected by a disruption in their system, the factor would be $2r/n = 2(5)/2 = 10{:}2$ or 5.

The two scores applied against the overall score indicate abilities to be resilient. This would feed up from tactical levels until an understanding of the operational levels and, finally, the strategic levels was possible.

Table 12.13Q   Qualitative Rating Scale

Weight	Description of Weighed Element
5	Disruption in the transportation system has an immediate, direct impact on the ability to maintain baseline services into the area and to accomplish the goals associated with the interdependency.
4	Disruption in the transportation system has a direct impact on the ability to maintain baseline services into the area and to accomplish the goals associated with the interdependency.
3	Disruption in the transportation system has an indirect impact that will gradually affect the ability to deliver the baseline services into the area and to accomplish the goals associated with the interdependency.
2	Disruption in the transportation system has an indirect impact that will cause a reduction in services provided but that will not affect the baseline performance required to maintain the interdependency.
1	Disruption in the transportation system has no bearing on the ability for the interdependency maintained.

12.15 Fragility Factor

Aligning each of these scalars across in such a way that a sum total of the results is generated can yield an initial, albeit crude, assessment of the fragility within the system. Aligning this approach with the mathematical models associated with the fragmentation and eventual dissolution of networks provides a much more comprehensive tool in terms of being able to arrive at the probability of failure within various levels of the system. This could be represented as follows:

Fragility Factor = (Fragility Score) × (Geography + Outside Reliance + Maintenance + Interdependencies) × (Prevalence to Single Points of Failure)

12.16 Relating to Resiliency and Redundancy

Essentially, within the transportation system fragility works in opposition to resiliency and redundancy by making the system brittle and unable to respond to the myriad disruptions that occur naturally throughout the system or that can happen for other reasons. At the tactical level, this involves the inability to restore and recover processes that maintain capacity within the system. Depending on the local service provided, this can also lead to disruptions throughout the system. At the operational and strategic levels, this might gradually involve erosion of trust in resiliency and redundancy levels of the system and the ability to maintain alternatives to work through or avoid disruptions within the system.

Thus, fragility flows from tactical to operational and to strategic levels and back again in a cycle requiring a clear and timely flow of useful information. One challenge involves the pace at which this information would have to flow to maintain that level of usefulness.

12.17 Fragility and the Path of Least Resistance

How disruption moves into various levels of the organization has been discussed already, but it is important to note that this disruption will likely manifest itself in areas that exhibit higher fragility scores within the system. This is not to say that other systems cannot be affected at all. If those systems have addressed many of the fragility issues identified, they will have identified the need for resiliency, redundancy, and robustness in their systems and have taken reasonable steps to prevent, detect, respond to, and recover from fragility detected within their systems. They will have reinforced this with the ability to identify, map, and predict any cascading failures within the system.

The flow disruption within the system will generally manifest itself in three ways. The first and immediate flow involves the shifting within the transportation system of demand that results from the disruption. This may involve the rerouting of traffic or traffic caught in the system. The second involves a loss of infrastructure through physical and economic pressures. Where the infrastructure is disrupted and the loss pushes the organization below a sustainable level, there is a risk that the infrastructure's capacity will be lost. The third follows the interdependencies between the various sectors that will be forced into a period of adjustment both in the immediate sense resulting from the disruption and the longer-term adjustments resulting from the loss of infrastructure.

12.18 Mean Time between Business Failure (MTBBF)

Fragility contributes a core environmental input into the MTBBF by identifying the level of instability within the environment. Businesses that operate within fragile environments will be at risk of a significantly lower MTBBF than those that are working within relatively resilient and robust systems.

This conclusion is based on the link between risks associated with loss of availability and with the loss of operational capability. As fragility increases, various levels of risk associated with the loss of inputs (from both internal and external sources) also increases. On one hand, this is based on the nature of the impact associated with the fragile aspects of the system, such as a lack of redundancy. On the other hand, it is also based on reasonable conclusions that as demands escalate incrementally on an increasingly unstable system, the likelihood of that system failing also rises. An increase in either of these two factors would, by definition, lead to an increase in risk.

This may also factor more heavily in emerging or immature markets in which margins of profit have not yet been sufficient to guarantee reserves in case of failure.

12.19 Mean Time between Market Failure (MTBMF)

Initially, emerging markets and the potential for primary and secondary demand leading to reasonable profit will draw the transportation system into creating local networks of nodes and conduits to meet that demand. Circumstances shifting may cause an adequate stress on those demands so that the network itself begins to allow to fragility to enter the system to preserve the anticipated levels of gain. This fragility will gradually increase until such a point as it becomes insurmountable unless revolutionary changes are made. Over time, the network will begin to decay, regardless of the best efforts of the participants within the network.

MTBMF refers to the failure of networks within that economic area. These failures are caused by organizations that function within interdependent networks falling prey to fragility-based issues due to their unwillingness or inability to incorporate the necessary resiliency, redundancy, or robustness in their individual systems. It should be noted that these three factors (i.e., resiliency, redundancy, and robustness) would require increasing levels of effort as fragility is allowed to creep into the system. As these entities begin to fail, their failure causes impacts within the interdependent network in a manner that is tied to both the fragility of the system at the operational level and the ability of the operational level to detect, respond to, and recover from the indicators and warnings of the impending failure.

This can also be present in mature markets that fail to identify their vulnerabilities that travel along interdependencies and fail to develop the necessary communities of interest and defensive networks. These communities are vital in maintaining the ability to identify, contain, and recover from the disruptions moving along the interdependencies. By failing to overcome management and organizational inertia, the system essentially isolates itself and allows itself to be unsuccessful within the network. As these breakdowns eventually lead to the fragmentation and ultimate collapse of the overall interdependent networks reflected by the market, the demands within the transportation system shift and cause a failing of the nodes and conduits that can no longer be supported within the system because of the loss of primary and secondary demands, making the efforts economically unviable.

12.20 Persistent Fragility Leading to System Revolution

Where systems fail to adapt, management may choose to adopt a revolutionary approach as opposed to ensuring that a learning system is established. This revolutionary approach becomes apparent as management crosses a threshold in its risk tolerance, particularly when aligned with risks of losses from similar events. Revolution may begin at any of the three levels. At the tactical level, it can be through actions undermining management to the point that management becomes ineffective. At the operational level, it can be through intolerance to the presence of risk or through desperation to put an end to a repeated pattern of events. Strategic levels of the organization may demand the action. Finally, where the impact is seen as being intolerable to society, the regulatory oversight may force that certain measures be put in place.

These revolutions, though sometimes useful, put the organization at significant risk of disruption as they tend to attempt to shift entire paradigms within the organization. These shifts have significant impacts on the personnel in terms of their understanding of procedures, their training, or their ability to manage risk within their day-to-day routines. Consequently, the organization may be at an even higher level of risk.

12.21 Management of Fragility

The management of fragility becomes an exercise similar to risk management. Some of the questions that management will need to address are these:

- How much fragility exists in the system, and is my management or leadership comfortable with that level of fragility?
- What vulnerabilities must I reduce or safeguards must I establish to bring that fragility back to acceptable levels?
- How does this fragility relate to the risks faced by my organization?

These questions align with previous figures outlined earlier in this chapter and can serve as a guide from which we may begin addressing the fragility issue.

These questions directly align with risk management in terms of the loss of availability of the system and can arguably provide a contributing factor to likelihood or probability in calculating that risk. By reducing fragility in the system, one is actually affecting means, opportunity, and intent by removing the opportunity to cause catastrophic impacts within the system that can lead to cascading failures or similar circumstances.

12.22 Relating to Prevention, Detection, Response, and Recovery

The doctrine of prevention, detection, response, and recovery can be reentered into the realm of fragility management. These four steps operate in the same manner as the traditional steps within security management or in emergency preparedness. The key will reside in the balancing of resources in the long term to prevent conditions in which the fragility of the system necessitates a revolutionary approach to the management of the system, putting significant strains on the overall system and leading to potential disruptions during the coordination process.

12.23 Transportation System Security, Risk, and Fragility

Securing the transportation system is an exercise in understanding the balance among infrastructure, capacity, risk, and fragility across tactical, operational, and strategic levels. Understanding the importance of infrastructure resides not in its dollar value but in the combination of that dollar value and its capacity that it delivers to the system. We can say that this provides one key to four locks. Understanding what puts that infrastructure at risk through a complex series of inputs and how that risk can move between the tactical, operational, and into the strategic levels provides the key to the second lock. Understanding the fragility of the system and the fact that risk and fragility, though not the same, operate in parallel with each other across an entire organization provide a significant step toward understanding that in its purest form fragility requires a significant portion of the organization to manage and respond to impacts and risks. Finally, knowing how this impact affects both the real and potential capacity of the system releases the fourth and final lock. These four locks provide a framework within which one can start to approach transportation system security within the critical infrastructure protection context.

12.24 Questions

1. Describe, using human physiology as an example, of how either MTBBF or MTBMF definitions could apply and would be characterized. At what point would the human body cease to function? What might be some of the extenuating circumstances in which the human body might continue to be alive, and how would that apply to the analogy?
2. Can you think of any additional elements that currently exist today that would play a vital or integral part in determining *fragility*? If yes,

please describe how these elements are considered vital in the overall framework defined?

3. Based on what you have read about the definition of *fragility*, how do you think it applies to your organization? Do you think your organization has a level of fragility, and if so, what would you do to reduce it?

4. What is the difference between *fragility* and *mean time between business failure*? How are the two terms similar? If possible, describe how they might be different from each other.

5. What are the differences, if any, between *mean time before business failure* and *mean time between business failure*?

Notes

1. *Spoofing* is "the interception, alteration, and retransmission of a cipher signal or data in such a way as to mislead the recipient" or "an attempt to gain access to an AIS by posing as an authorized user." Institute for Telecommunication Sciences, "Spoofing," *Telecom Glossary 2000*, http://www.its.bldrdoc.gov/fs-1037/ dir-034/_5049.htm. An alternate definition is "a technique used by hackers to access computer systems by modifying packet headers to make them appear to have originated from a trusted port" or "the practice of falsifying an e-mail header to make it appear as though it originated from a different address." Institute for Telecommunication Sciences, "IP Spoofing," *Telecom Glossary 2000*, http://www.its.bldrdoc.gov/projects/devglossary/_ip_spoofing.html.

2. An alternative term used to spoof a computer transmission signal in which the potential attacker piggybacks a payload signal on top of a returning signal to the attempted would-be "victim." This method is often used in spoofing techniques specifically relating to network security.

3. Holling, C.S. (1985). Resilience of ecosystems: Local surprise and global change. In: Malone, T.F. and Roederer, J.G. (Eds.), *Global Change*, Cambridge University Press, Cambridge, MA, pp. 228–269. (rs.resalliance.org/author/buzz-holling).

Sample Memorandum of Understanding between The Radio Amateurs of Canada Inc. and The Canadian Red Cross Society[1]

Memorandum of Understanding between The Radio Amateurs of Canada Inc. and The Canadian Red Cross Society

The Canadian Red Cross Society recognizes that the Radio Amateurs of Canada Inc., because of its excellent geographical coverage, can render valuable aid in maintaining the continuity of communications during disasters and emergencies when normal communications facilities are disrupted or overloaded.

The Radio Amateurs of Canada Inc. recognizes The Canadian Red Cross Society as an agency that provides assistance to individuals and families affected by disasters in Canada and around the world through the International Committee of the Red Cross and the International Federation of Red Cross and Red Crescent Societies.

Whenever there is a disaster or an emergency requiring the use of radio communications facilities, the Radio Amateurs of Canada Inc. agrees to provide, whenever and wherever possible:

1. The alerting and mobilization of volunteer emergency communications personnel and equipment in accordance with a pre-determined plan.
2. The establishment and maintenance of fixed, mobile and portable emergency communications facilities for local radio coverage and point-to-point contact between Red Cross and various locations, as required; and
3. Adequate provision of service for the duration of the emergency or until substantial regular communications are restored and stand down is ordered by Red Cross Emergency Services.

---

[1] Both authors wish to thank the executive staff of the Radio Amateurs of Canada (RAC) for allowing us to use their memorandum of understanding (MOU) as an example of a well-written one between two volunteer organizations. A sample MOU is provided—untouched—in its entirety.

This Memorandum of Understanding will remain in effect provided that either party may terminate this Memorandum of Understanding by giving the other party three months notice in writing of its intention to so terminate.

Further details concerning the method of cooperation are outlined in Appendix A. Information on the organization of The Canadian Red Cross Society and the Radio Amateurs of Canada Inc. is attached as Appendix B.

Signed by:
President
Radio Amateurs of Canada Inc.
National Director, Field Operations
The Canadian Red Cross Society
April 28, 1994

APPENDIX A

Guidelines for Cooperation

1. Through its executive level, Radio Amateurs of Canada Inc. will maintain liaison with The Canadian Red Cross Society's Emergency Services in order that there may be the closest possible cooperation in emergency communications planning and the coordination of radio communication facilities for disaster relief operations.
2. Red Cross Divisions, Regions and Branches are encouraged to invite one or more members of the amateur radio community to serve as Red Cross volunteers for emergency preparedness and relief.
3. Personnel of the Radio Amateurs of Canada Inc. are eligible for reimbursement by Red Cross for reasonable out-of-pocket and travelling expenses while conducting approved business on behalf of the Society.
4. Detailed operating plans for the full utilization of the communications facilities of the Amateur Radio Emergency Service should be developed by the local Red Cross in cooperation with the Radio Amateurs of Canada Inc.'s local Emergency Coordinator.
5. The Canadian Red Cross Society will recommend to its Divisions that membership on disaster preparedness and relief committees include representation from the appropriate officials of the Radio Amateurs of Canada Inc.
6. The Canadian Red Cross Society will furnish Divisions with copies of this statement of understanding and the Radio Amateurs of Canada Inc. will similarly furnish copies to its Field Officials.

OCTOBER 25, 1993

APPENDIX B

Organization of The Canadian Red Cross Society

1. The National Headquarters of The Canadian Red Cross Society is located in Ottawa. For administrative purposes, Canada is divided into ten Divisions with each Division having jurisdiction within its own Province. Divisional Offices are located in the following cities: Burnaby, B.C.; Calgary, Alberta; Regina, Saskatchewan; Winnipeg, Manitoba; Mississauga, Ontario; Verdun, Quebec; Saint John, New Brunswick; Halifax, Nova Scotia; Charlottetown, Prince Edward Island; and St. John's, Newfoundland. The B.C./Yukon Division and the Alberta/NWT Divisions are responsible for Red Cross operations in their respective Territory.

2. Regions and Branches are the local units within each Division of The Canadian Red Cross Society. These units are responsible for all local activities of the Red Cross within its territory, subject to the policies and regulations of the divisional and national organization.

3. Each Region and Branch is responsible for developing an Emergency Services Committee of the best qualified volunteers available. This Committee studies the disaster hazards of the territory and surveys local resources for personnel, equipment and supplies, including transportation and emergency communication facilities, that are available for disaster relief. It also formulates cooperative plans and procedures with local governmental agencies, private and other volunteer organizations for carrying on relief operations should a disaster occur.

Organization of The Radio Amateurs of Canada Inc.

4. The American Radio Relay League Inc., (ARRL) was founded in 1914 to encourage and support every aspect of amateur radio. The ARRL became a bi-national organization in 1920 with the formation of the Canadian Division and Canadian membership.

5. The Canadian Division was known in Canada as the Canadian Radio Relay League Inc., (CRRL) giving it a distinctly national entity. The CRRL elected officers were charged with policy administration as established by their Executive Committees and Board of Directors. On May 2, 1993 the Canadian Amateur Radio Federation and the Canadian Radio Relay League Inc. ceased operation and merged together on that day to form the Radio Amateurs of Canada Inc. This agreement will then continue in force with Radio Amateurs of Canada Inc. who will carry on with the Field Services Organization.

6. The Radio Amateurs of Canada Inc.'s Field Services Organization operations are administered by the Field Services Manager through elected

Section Managers (SM). Canada is divided into seven sections: British Columbia-Yukon, Alberta and the North-West Territories, Saskatchewan, Manitoba, Ontario, Quebec and the Atlantic Provinces, New Brunswick, Nova Scotia, Prince Edward Island and Newfoundland forming the Maritime/Newfoundland Section.

7. The Radio Amateurs of Canada Inc. sponsored Amateur Radio Emergency Service (ARES) consists of two branches: the Amateur Radio Emergency Service (ARES) and the National Traffic System (NTS). Both branches work together, are supported by thousands of licensed radio amateurs and are under the jurisdiction of their Section Manager.

8. ARES—Amateur Radio Emergency Service. The ARES is an organization of licensed radio amateurs who have voluntarily registered their qualifications and equipment with the Radio Amateurs of Canada Inc. for communication duty when disaster strikes. It is supported and directed by Radio Amateur of Canada Inc. appointees. The leading provincial ARES official is the Section Emergency Coordinator (SEC) who appoints individual Emergency Coordinators (ECs) and District Emergency Coordinators (DECs) across the province to assist locally. It should be noted that membership in ARES is not restricted to members of RAC.

9. NTS—National Traffic System. The NTS compliments the ARES and functions daily in the handling of medium and long distance formal message traffic and whose network operations can be stepped up to meet the needs of an emergency situation. The leading NTS official is the Section Traffic Manager (STM) who is assisted by carefully trained and selected Net Managers (NMs). Traffic nets link with other nets throughout North America and South America, the Caribbean and Australia and operate every day and night of the year. Further-training, tests and drills for the ARES and NTS members maintain a disciplined readiness in providing emergency communications.

10. Radio Amateurs of Canada Inc.'s Section officials (SM), (SEC), (STM) work closely together daily as well as with the organization's Headquarters and/or Government officials as required during emergency situations.

Don Shropshire
February 11, 1994

Manager's Working Tool

This tool is intended to assist managers in looking at their organizations in terms of infrastructure assurance and capacity management. This involves becoming aware of the organization's strengths and weaknesses. Once this process is completed, a manager should be able to return to Chapter 12 and fill in the fragility scores for his or her organization. This will assist the tactical level in the provision of a baseline on which to improve. For the operational or strategic level, this will assist in being able to determine the level of trust that can be imparted to an organization with respect to its ability to assure performance. This chart should not be considered a substitute for sound professional judgment. A graphical representation of these tables is provided for flow continuity reasons.

Section 1: Product or Service Delivery

This series deals with your organization's ability to deliver its services. Remember to take an all-hazards approach that includes natural disasters, failures to receive key inputs, disruptions in utilities for a number of reasons, labor or similar disruptions, disruptions in the community, cyber attacks, crimes against the organization, and terrorism. Some of these will be far less likely to occur than others.

Series Done	Item
1	Does your organization have a clearly stated mission?
1Y	*Go to 2.*
1N	*What is the primary revenue generator or mandate of your organization?*
2	Have you identified a key/core/critical process that delivers that mission?
2Y	*Go to 3.*
2N	*Consult with operations to identify that key process. Then go to 3.*
3	Have you identified how success is measured for that core process (e.g., a certain number in a given time, done within a certain cost)?
3Y	*Identify the level for desired, break-even point, and unviable level. Go to 4.*
3N	*Identify with financial and operations the levels involved. Go to 4.*

(Continued)

Series Done	Item
4	Have you identified which people contribute to the process?
4Y	*If a person were not available, could the process continue, or how much would it fall behind given the goals outlined in question 3? Go to 5.*
4N	*Identify people involved in the process. Are there any that, if absent, would cause the process to fail, or how much would it slow if they were not available? Go to 5.*
5	Have you identified which assets or objects contribute to the process?
5Y	*If an object was not in service or was not functioning up to standard, what would the effect on the process be? Go to 6.*
5N	*Identify objects and equipment that contribute to the process, and identify how much the process would be affected if they did not function as intended or not at all? Go to 6.*
6	Have you identified which facilities contribute to the process?
6Y	*What would the effect on the process be if the facility could not be used? Go to 7.*
6N	*In consultation with operations, identify how the process would be affected if the facility were not available. Go to 7.*
7	Have you identified what information is needed?
7Y	*If that information were not available or could not be trusted, what effect would it have on the process?*
7N	*Discuss with operations what information is needed and what the effect would be if it were not available or inaccurate.*
8	Have you identified any activities that are needed to perform the operation?
8Y	*What would the effect be if that activity was not performed or was not performed correctly?*
8N	*With operations, identify what activities have to take place. Include inspections, calibrations, work, and shut-down procedures necessary to maintain the level of performance.*
9	Where you answered "NO" to any of the above, identify how long it would take to get things back to normal and the process back on track. Also, identify the maximum time something could be down before the disruption is considered intolerable or completely unacceptable.
10	Have you identified processes either that support your core business function or that allow your business to remain in operation (e.g., pay)?
10Y	*Repeat questions 3 through 9 for these processes, documenting the persons, objects, facilities, information, and activities involved.*
10N	*Repeat questions 2N through 9 for these processes, documenting the persons, objects, facilities, information, and activities involved.*
11	When looking at the list of persons, objects, facilities, information, or activities involved, have you identified alternatives or backup plans that would allow for the same results to be achieved or for the process to continue at an acceptable rate?

(Continued)

Series Done	Item
11Y	*Identify these in terms of their support to the main process or other processes and under what conditions they would be activated or called upon.*
11N	*For each person, object, facility, information, or activity involved, attempt to identify an alternative should it fail. This may not always be possible; if not possible, these items should be clearly identified to management as single points of failure.*
12	For each of the persons, assets, facilities, information sets, and activities, do you have some physical, procedural, or psychological procedure in place to prevent its failure, detect that it might fail, respond to it if does fail, or recover from its failure?
12Y	*Document these or make reference to where these are documented in a single location. If possible, try to ensure that you use existing measures as much as possible, reinforcing them as opposed to creating new ones.*
12N	*The goal here is ideally not only to prevent its failure but also to be able to detect and respond to its potential failure before the failure can take place. This may not always be possible, so a recovery plan (minimum time) can also be useful for each.*
13	Are these plans well documented, with a person responsible for ensuring that they are kept up to date?
13Y	*This is a very important function. A record should be kept with respect to the last update to determine if plans need to be reviewed and so forth. If possible, this should be done by someone who knows when the operations or risks to the operations have changed within the organization.*
13N	*Identify a manager or responsible person who will keep this up to date and recorded.*
14	Have you identified, for example, any stores, supplies, or services that originate from outside the organization that you require to keep the process moving at an acceptable level?
14Y	*Ensure that you have contracts or other legally binding agreements that guarantee that those items will be delivered on time, at the right location, and in the right condition.*
14N	*Identify these items and their suppliers. Ensure that you have contracts or other legally binding agreements that guarantee that they will be delivered on time, at the right location, and in the right condition.*
15	Do any of the items in 14 acts as a single point of failure? This involves the product, goods, or service coming either from a single source or through a single channel in such a way that if it failed, you would be unable to continue the process.
15Y	*Identify alternatives or plans to recover from the losses as quickly as possible.*
15N	*Ensure that arrangements are formalized so that the backup plans and so forth are as stable as possible.*

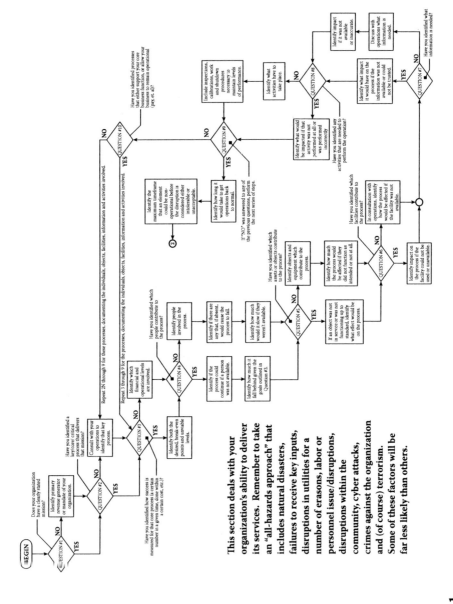

This section deals with your organization's ability to deliver its services. Remember to take an "all-hazards approach" that includes natural disasters, failures to receive key inputs, disruptions in utilities for a number of erasons, labor or personnel issue/disruptions, disruptions within the community, cyber attacks, crimes against the organization and (of course) terrorism. Some of these factors will be far less likely than others.

Figure B.1

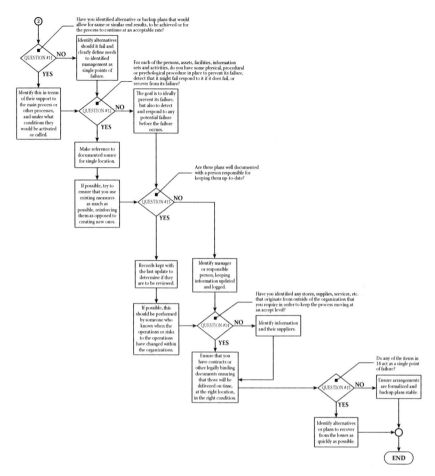

Figure B.2

Section 2: Geography and Community Building

This series deals with the action of the organization within the community with the goal of establishing trusted communities that can assist in the assurance of infrastructure. Ensuring this capacity can significantly reduce the overall perceived risk as the community and organization come together in mutual self-interest to the benefit of both.

Series Done	Item
16	Does your organization consist of more than one office or installation?
16Y	*Identify a person responsible for each office or installation that will act as a point of contact. Remember, if this person becomes responsible for tasks and such, resources may be required in support of those tasks.*
16N	*Ensure that the office has a stable point of contact identified for the office.*
17	Does your organization own or rely on infrastructure that is remote or removed from your control?
17Y	*Identify the organization responsible, and identify how they prevent disruption, detect the possibility of disruption, respond to indicators of disruption, or recover from disruption. You may consider entering into an arrangement requiring them to assure the use of the infrastructure or provide a suitable alternative at no additional cost.*
17N	*Ensure that you have the ability to prevent the loss of use of that infrastructure, detect potential disruptions, respond to indicators of disruption, or recover from disruptions.*
18	Does your organization use networks of routes or similar infrastructure to deliver its goods or services?
18Y	*Identify the organization responsible for those routes, and identify how they prevent disruption, detect the possibility of disruption, respond to disruption, and recover from disruption. Attempt to establish a line of communication that will identify disruptions to your organization as quickly as possible, and establish a procedure on how to plan around them.*
18N	*Verify that you do not rely on this kind of network for inputs to the process, and if you do repeat the steps for 18Y.*
19	Does your organization rely on single points of access or egress that could be disrupted or be made unavailable, causing disruptions within the organization?
19Y	*If possible, identify or establish alternative routes that can be activated. These routes should be secured—understanding the vulnerability associated with a potential point of access—and maintained in case of need.*
19N	*If possible, identify routes in order of greatest need, ensuring the availability of those with highest traffic first and then working backward.*
20	Have you established ties with the community (e.g., citizens, government, major organizations) and convinced them of the need to provide the infrastructure around your organization?
20Y	*Reinforce the value of the infrastructure to the community, being careful not to overstate it.*
20N	*Establish a sense in the community that they have a vested interest in your organization being up and running.*
21	Have you identified methods of communication, within the communities, that can be used to broadcast the need for vigilance?

(Continued)

Series Done	Item
21Y	*This may involve communities across a range of interests, such as HAM operators willing to assist in communicating reliable information. Reinforce the value of their participation in assisting you in detecting indicators or warnings of potential disruption. Make certain this is linked to the value of the community so as not to isolate participants or informants.*
21N	*Identify leaders and how they communicate into their community. Certify that they are on your side, and indicate what to look for and what to do if they see it.*
22	Have you identified a method by which those outside of your immediate location can communicate the presence of potential disruptions to you?
22Y	*Reinforce the value of this activity, and make sure that the indicators and warnings being looked for are up to date. Test the lines of communication periodically.*
22N	*Establish a method by which your organization (or some group acting on your behalf) can be contacted on a 24-hour basis, and assure that the community is aware of what needs to be reported using that method.*
23	Have you identified, across your organization, a standardized method of reporting that a disruption is occurring?
23Y	*Guarantee that this method is up to date and that the data being communicated are clearly labeled in terms of their application and sensitivity.*
23N	*Establish a system that involves communicating disruptions, including the time, location, nature, estimated time of resolution, and steps being taken to address the disruption. Ensure that this system is used across the organization.*
24	Does your organization have plans to be able to chart routes or movement around disruptions or clearing the disruptions?
24Y	*Ensure that these plans are up to date and are exercised from time to time. Ensure that all different management levels are involved: local (tactical), regional (operational), and organizational (strategic).*
24N	*Develop plans to respond to notification of disruption that involve the ability to clear the disruption or chart movement around it. Note the capacity of alternatives to deliver the same services, particularly where costs are comparable.*
25	Does the organization track disruptions across the overall organization, including their causes and effective responses to them?
25Y	*Determine why certain responses worked, and communicate this across the organization. Conversely, to prevent future recurrence determine why some responses did not work.*
25N	*Identify a method by which you can identify strategic routes and movement that will achieve those goals. Attempt to identify as many alternatives as possible. For the most part, you may wish to consider entering into contingency agreements.*

(Continued)

Series Done	Item
26	Are you involved with local community groups that can either communicate for or inform you of disruptions in the system in terms of awareness, familiarization, or even training programs?
26Y	*Reinforce lines of communication and familiarization, ensuring that information (including contact information for primary and secondary personnel) is up to date and that lines of communication are checked periodically.*
26N	*Identify local organizations or committees involved in law enforcement, intelligence, commerce, or similar activities (e.g., port security, airport security, tourism and commerce, trucking), and become involved in them. Establish lines of communication to identify areas of concern but also the ability to detect disruptions in the system through trusted sources.*
27	Do you access or research trends associated with changes in the operating environment or threat environment of your organization?
27Y	*Ensure that you have links to local, regional, and national unclassified threat assessments identified and communicated into operations and management through the organization. Build awareness of the value of this information in terms of business continuity.*
27N	*Note the daily infrastructure reports from the U.S. Federal Emergency Management Association (FEMA) and Public Safety (Canada) and similar reports for information technology. Ensure that these are communicated across operations, and consider preparing a weekly briefing based on a compilation of them (1–2 minutes or 5 slides) that can be used by management.*
28	Have you identified primary, secondary, and other abilities to communicate voice, fax, data, or other information?
28Y	*Ensure that appropriate agreements are in place. Consider providing assurance that senior management or operations personnel are identified for primary restoration of services.*
28N	*Ensure that lines of communication are established linking your organization. Be creative, noting the kind of information that you would need to reestablish operations, and be sure to brief management and communicate those requirements (in general terms and not in a way to expose vulnerabilities) into those lines of communication.*

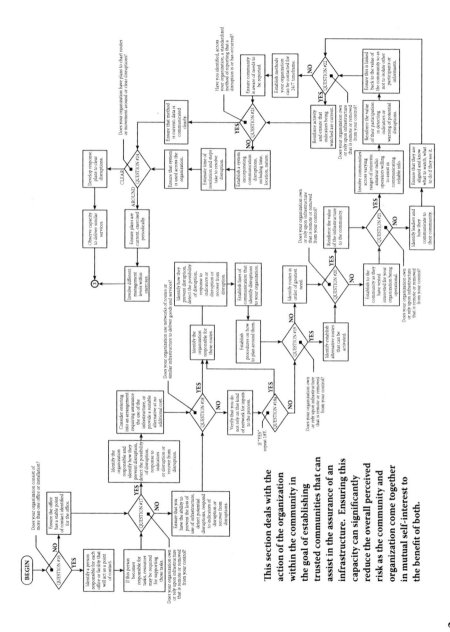

This section deals with the action of the organization within the community in the goal of establishing trusted communities that can assist in the assurance of an infrastructure. Ensuring this capacity can significantly reduce the overall perceived risk as the community and organization come together in mutual self-interest to the benefit of both.

Figure B.3

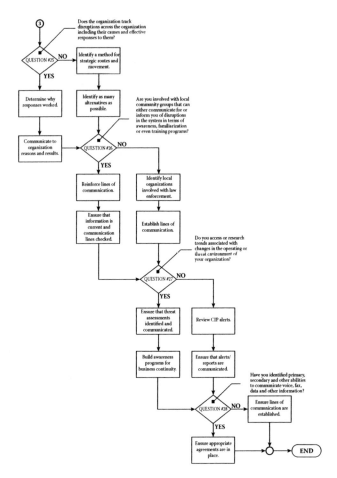

Figure B.4

Section 3: Data Categorization and Information Management

The ability to categorize and sort data, to understand the value of information, and to generate intelligence is crucial to an organization's ability to establish a dynamic learning system. These questions will assist the manager in determining how refined this capacity is within the organization.

Series Done	Item
29	Does you organization use internal information management systems that generate their own data?
29Y	*Ensure that these systems operate in a stable environment, including the ability to establish resiliency (short down times) or redundancy (backup systems).*
29N	*Identify outside systems, and certify that agreements are in place to ensure the availability of those systems so that they perform at acceptable levels.*
30	Does information originated within your organization have a level of sensitivity or value?
30Y	*Identify whether it is the confidentiality, integrity, or availability of the information that is of importance, and take steps that will preserve that quality based on the order of value.*
30N	*Verify with management and the holders of the information that the information could not be used should it fall into the hands of a competitor or potential attacker. Revisit 30 if necessary.*
31	Does the information generated clearly indicate the level of sensitivity?
31Y	*Ensure that personnel understand how the information is labeled or how to identify sensitive information. Also, verify that they know how to handle it (e.g., storage, transport, mailing).*
31N	*With operations and network/data custodians, name a system that can be consistently applied across the organization that will identify information that is sensitive. Ensure that if this done data are protected from unauthorized disclosure (remember, identifying it as sensitive may make it easier to target). Guarantee that the results are communicated as per 31Y.*
32	Is information sorted and then identified in terms of its core functions, values, and organizations?
32Y	*Check that information is protected from unauthorized disclosure.*
32N	*With network or data custodians, identify a system that tags information in terms of its core contribution to the organization, values (confidentiality, integrity, or availability), and functions.*
33	Does your organization relate data and information to each other; that is, does your organization identify data as being unprocessed information?
33Y	*Identify the key or capable technical authority that can assist in the configuration of data flows in support of the movement of information.*
33N	*Information provides an input into the process that has value. Consider data in terms of the best structure in support of that input.*
34	Does your system manipulate data for communication based on their confidentiality, integrity, or availability?

(Continued)

Series Done	Item
34Y	*Carefully document the manipulation, and relate it back to the information and value of the information to the input to the process.*
34N	*Consider manipulating data based on confidentiality, integrity, availability. Where data are required to be kept confidential, consider dividing them up and communicating them across many channels so that an intercepted chunk cannot be used. Where integrity is involved, consider minimizing the number of divisions and configuring the system to make certain that minimal disassembly and assembly occurs. Where availability is an issue, consider the movement of data across three or numerous channels to take care that the information arrives at its destination. Note, where two or more data streams come together, incorporate error or conflict resolution.*
35	Does your system maintain the ability to back up data, information, and then intelligence? The first two may be accomplished with backup systems whereas the latter leans toward the ability to maintain redundant systems to preserve the processing of that information.
35Y	*Be sure to look at systems in terms of resiliency, redundancy, and availability.*
35N	*Consider the implementation of backup or redundant capabilities as appropriate to assuring the availability of these.*

Section 4: Establish a Learning System

Given the previous section, the next step is to use those components in such a way that the system is able to learn from both failure and success. This involves identifying and reinforcing positive behavior while discouraging negative behavior.

Series Done	Item
36	Does the organization test its plans?
36Y	*Record successes and failures, and determine the causes for each. Communicate these into the next level of the organization.*
36N	*Testing plans (drills and exercises) is crucial to ensuring that the measures put in place will function.*
37	Does the organization conduct tests such as investigations and audits?
37Y	*Where systems appear to fail or allow disruptions, communicate the reasons for failure (or success) into the organization.*
37N	*Investigations and audits provide structures for both performance measurement and accountability. These should be integrated into basic management structures.*
38	Does your organization evaluate failures against the information contained in the risk assessment?

(Continued)

Series Done	Item
38Y	*Verify that such information is assessed and records are kept of the assessments made.*
38N	*When failures in the system occur, two evaluations should take place. Did safeguards perform as intended; why or why not? Do vulnerabilities exist to the same extent as identified? When incorporating these back into the base risk assessment (as opposed to amending a plan), gaps in the system may become apparent in apparently unrelated areas.*
39	Does your organization have a system for recording successes and failures at all levels in the organization?
39Y	*Check that the system is protected from unauthorized disclosure but available to those responsible for monitoring the effectiveness of the system across the organization.*
39N	*Establish a system within the framework of acceptable corporate culture to assist in preventing waste from unnecessary recurrences of failure and to maximize the potential of success.*
40	Does your organization participate in outside groups that focus on the improvement of the organizations' systems?
40Y	*Certify that communication is involved to prevent the unauthorized (inadvertent) disclosure of sensitive information that could be used against the organization (e.g., planning). Reinforce methods that allow the full organization to participate.*
40N	*Involvement in these organizations can prove useful in terms of gaining valuable insight into challenges but can also pose a risk in terms of providing information to competitors. Information sharing arrangements should be carefully defined and controlled.*

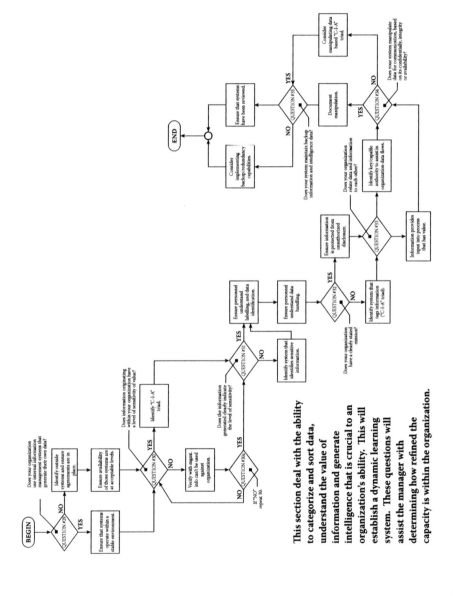

This section deal with the ability to categorize and sort data, understand the value of information and generate intelligence that is crucial to an organization's ability. This will establish a dynamic learning system. These questions will assist the manager with determining how refined the capacity is within the organization.

Figure B.5

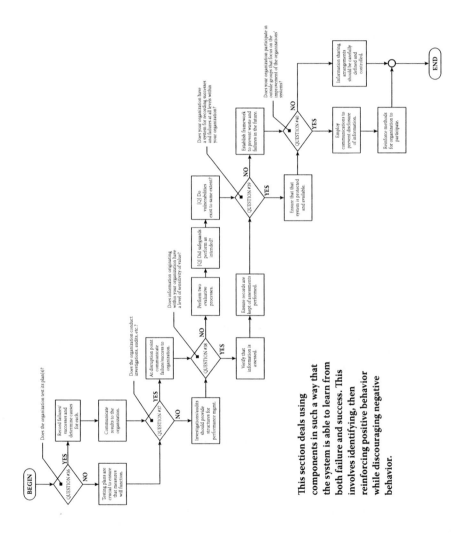

This section deals using components in such a way that the system is able to learn from both failure and success. This involves identifying, then reinforcing positive behavior while discouraging negative behavior.

Figure B.6

Section 5: Maintenance and Sustainability

This section refers to the ability to preserve management support while also keeping costs to a reasonable level and preventing the need for catastrophic levels of funding to fix vulnerabilities.

Series Done	Item
41	Does your organization maintain a preventive maintenance cycle?
41Y	*Linking this cycle with the understanding of vulnerabilities in the equipment can identify potential risks not generally apparent to nonexperts.*
41N	*Establishing preventive maintenance cycles not only can reduce costs but also can raise awareness of operational issues. Larger, unexpected impacts can be avoided by catching glitches early. Including operator feedback can also be of significant value.*
42	Do your suppliers maintain preventive maintenance cycles or maintain records of actions taken to prevent disruptions due to the failure of infrastructure?
42Y	*Establishing a routine of verifying this can assist in demonstrating a sense of accountability in your suppliers.*
42N	*Where these steps are not being taken, you may be at risk of a loss of an input due to an unexpected failure in their infrastructure. Establishing legally binding agreements to preserve levels of performance or somehow reducing the impact (by threat of penalties) should be considered at a minimum.*
43	Does your organization operate an awareness or training program that ties these activities back into the value of the business?
43Y	*This can assist in stabilizing the program, particularly if you can show the value these activities provide by contributing to the organization's success.*
43N	*Lack of employee and personnel awareness of issues can be a critical vulnerability in the system.*
44	Does your organization promote positive reinforcement in addition to compliance?
44Y	*Positive reinforcement, particularly when supported by senior management and concrete value for effort, can significantly increase the buy-in to a program. It can also support partnering by providing a balance to the impression that these programs are compliance based only.*
44N	*Programs that rely on compliance and sanctions only run the risk of breaking partnerships as parts of the organization attempt to fool those coming out to check on the status of things. This can leave undetected vulnerabilities in the system.*

As a rule of thumb, the questions here are tailored so that where Y answers are returned the possibility of fragility existing in the system drops.

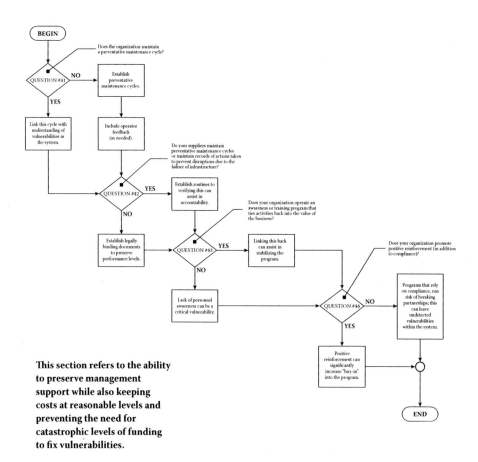

This section refers to the ability to preserve management support while also keeping costs at reasonable levels and preventing the need for catastrophic levels of funding to fix vulnerabilities.

Figure B.7

Index

G

H

I

U

V

W

Z